The Reflection of Life

International Federation for Systems Research International Series on Systems Science and Engineering

IFSR was established "to stimulate all activities associated with the scientific study of systems and to coordinate such activities at international level." The aim of this series is to stimulate publication of high-quality monographs and textbooks on various topics of systems science and engineering. This series complements the Federations other publications.

A Continuation Order Plan is available for this series. A continuation order will bring delivery of each new volume immediately upon publication. Volumes are billed only upon actual shipment. For further information please contact the publisher.

Volumes 1–6 were published by Pergamon Press

For further volumes:
http://www.springer.com/series/6104

A.H. Louie

The Reflection of Life

Functional Entailment and Imminence
in Relational Biology

 Springer

A.H. Louie
Ottawa, Ontario
Canada

ISSN 1574-0463
ISBN 978-1-4899-8905-5   ISBN 978-1-4614-6928-5 (eBook)
DOI 10.1007/978-1-4614-6928-5
Springer New York Heidelberg Dordrecht London

Mathematics Subject Classification (2010): 92B99

Springer is part of Springer Science+Business Media (www.springer.com)

To the *rami*[fications] of
my *lignum vitae* and *arbor scientiae,*
my genealogical and academic progenies:

vos palmites estis
ut fructum plurimum adferatis

There are more things in heaven and earth, Horatio,
Than are dreamt of in our philosophy.

— William Shakespeare (*c.* 1600)
The Tragedie of Hamlet, Prince of Denmarke
Act I, scene v (*First Folio* text, 1623)

Praefatio
Le reflet de la vie

J'ai toujours préféré le reflet de la vie à la vie elle-même.

[I have always preferred the reflection of life to life itself.]

— François Truffaut (1970)
Téléciné, No. 160
(«Spécial Truffaut», mars 1970)

Welcome to the continuation of our exploratory journey in relational biology! My previous book

More Than Life Itself: A Synthetic Continuation in Relational Biology

was published in 2009. It dealt mainly with the epistemology of life. In its Chapter 13, Ontogenic Vignettes, I briefly mentioned several topics that would be expanded elsewhere, in "my next book". This monograph you are now reading is that "elsewhere". It will deal with the ontogeny of life as well as how life evolves from the singular to the plural. This 'Opus II' of my epic on relational biology is thus a 'second image', hence 'reflection'.

The roots of the Latin word *reflectere* are *re* 'back' and *flectere* 'to bend'. In geometry, a 'reflection' (also spelt 'reflexion') is an isometric mapping from a Euclidean space to itself that has a hyperplane as the set of fixed points. When a point is reflected about an axis, for example, the point is 'bent back' to a symmetric position on the opposite side of the axis. A reflexive relation 'bends back' every element so to be related to itself. In physics, 'reflection' is the transition, 'bending back', of a wavefront at an interface between two different media so that the wavefront returns into the medium from which it originated. Metaphorically, the word 'reflection' can mean 'turning back one's thought on some subject', whence long and careful consideration, an indication, an account, or a description. 'Reflection' is a noun of action; it entails plurality. Any object may be the material cause of reflection and be bent back under a formal cause of reflective morphism. The efficient cause of reflection is the interaction of the to-be-reflected entity with its reflector (that which reflects), and the final cause is the genesis of the reflected output. Common reflected entities are light, heat, sound, and water waves, and—by extension—colour, image, thought, concept, and idea, thence verily exemplified in the sight and sound of *la Nouvelle Vague* that is above all 'human self-reflection'.

This *liber secundus* of my synthetic continuation in relational biology is, therefore, a 'reflection' in every literal and metaphoric sense of the word. Indeed, modelling, the representation of one system in another, is the art that is the ultimate revelatory reflection of life. This is why I have chosen to name this book *The Reflection of Life* (and, for me, the exceedingly *à propos* Truffaut quote clinches it). I nominate it thus, despite being fully aware that the title is somewhat generic and formulaic: the shelf of books entitled *The Y of X* is quite crowded. (Incidentally, *The Origin of Species* is not a fitting example here. Although this arguably most famous scientific publication is often referred to by this more declarative name, Charles Darwin's original 1859 title was the verbose *On the Origin of Species by Means of Natural Selection, or the Preservation of Favoured Races in the Struggle for Life*.) Even in my subject area of mathematical biology, the name *The Y of Life* is well represented; among them are, for example, Denis Noble's 2006 *The Music of Life* and Ian Stewart's 2011 *The Mathematics of Life* (both, I may add, excellent books). My rather specific subtitle for the book should, nevertheless, serve to distinguish it: I am reasonably certain (in the strong-limit sense of almost sure convergence), an infinitude of typing monkeys notwithstanding, that the very sequence of words *Functional Entailment and Imminence in Relational Biology* has not appeared in print elsewhere.

A main theorem in relational biology says:

> *A natural system is an organism*
> *if and only if it is closed to efficient causation.*

If such a central issue of what life is can be so succinctly defined, then why is relational biology not as well known as it deserves to be? It may be because category theory, the *lingua franca* of relational biology, is not a very accessible branch of mathematics; it is not uncommon for a university student graduating in mathematics not to have taken a course on the subject. It may also be true that many in the rest of the community of biologists at large were antagonistic towards the Rashevsky-Rosen school, perhaps not so much on petty personal(ity) conflicts than on points of philosophical difference.

We are not denying that an underlying material basis is needed and that *some* information on living systems may derive from their material bases. The real *nature* of living systems, however, is not conveyed by their material basis. Physicochemical structures do not dictate functions; physicochemical structures are manifestations of functions.

Many biologists are convinced that "biology is inherently messy", and some aggrandizers have even presumptuously spoken for all and proclaimed as a "conviction" of biologists that the actual complex behaviour of real organisms would be lost in simple even if elegant idealizations. They regard cells and organisms as machine-like systems, a metaphor that even today dominates biology. Even for those biologists that are not as blatantly reductionistic, they would still brand relational models "(over-)simplifications", and advocate (and advertise) the euphemistic "biologically realistic models" or "models of biological relevance". But what do "realistic" and "relevant" imply? Do they not implicitly remain the insistence that everything in biology must be explainable in terms of the underlying physicochemical materials? Contrariwise, from the standpoint of relational biology, machine-like systems are in fact simple; biological systems are complex precisely because their essence is lost when modelled as machines.

I may conjecture that this physicochemical bias has puritanical roots. Let me state that I am not referring to (capitalized) Puritanism that is the theological creed and social vision, but only to a debased, secularized, conservative form of (lower-case) puritanism, that of "anguished self-flagellation" and "suffering is purposeful". To wit, the slogan of many experimental biologists is that "real biologists" must "get their hands dirty", and that they must keep their "feet on the ground" (extolled from their *pieds-à-terre* in ivory towers; *cf.* [Rosen 2006] for an anecdote)! It is not that they do not appreciate that nature *itself* is beautiful; it is just that they feel the worthiness of an experimenter's *study* of nature ought somehow to be linked to the degree of messiness and dirtiness of the endeavour.

I wonder how people can appreciate the ontological beauty of nature but then insist on its epistemological ugliness.

Function dictates structure: relational biology begins with mathematical ideas and seeks realizations in natural systems. The Book of Nature is written in the language of mathematics. A theorist's conception of nature is based on *beauty*. I shall let G. H. Hardy, pure mathematician *par excellence*, have the last word:

> The mathematician's patterns, like the painter's or the poet's, must be *beautiful*; the ideas, like the colours or the words, must fit together in a harmonious way. Beauty is the first test: there is no permanent place in the world for ugly mathematics.

> — G. H. Hardy (1940)
> *A Mathematician's Apology*
> § 10

A. H. Louie
19 May, 2012

Nota bene
Prerequisites

The cast and crew of mathematical and biological characters in 'Opus I', my previous book *More Than Life Itself* [Louie 2009], include partially ordered sets, lattices, simulations, models, Aristotle's four causes, graphs, categories, simple and complex systems, anticipatory systems, and metabolism-repair [(M,R)-] systems. In this 'Opus II', my present book *The Reflection of Life*, I shall expand the cast and crew to employ set-valued mappings, adjacency matrices, random graphs, and interacting entailment networks. If the theme of Opus I is *one* (M,R)-system, then the theme of Opus II is *two* interacting (M,R)-systems.

Throughout this book I shall adopt the notation and terminology and draw upon results from *More Than Life Itself*. Since I shall be referring to that book many times, henceforth the canonical symbol *ML* will be used in its stead. In this present volume, when various topics are encountered, when appropriate I shall refer the reader to relevant passages in *ML* for further exploration; the notation '*ML*: m.n' refers to Section m.n (in Chapter m) of *ML*.

I assume the reader is already familiar with the premises of the Rashevsky-Rosen school of relational biology, as explicated in *ML*. In particular, I recursively enlist all the assumptions made in the Nota bene of *ML* (*pp*. xxiii–xxiv) and include them as prerequisites for continuing our journey in relational biology. The Exordium that follows next is a terse introduction to relational biology, but it is a précis, and not a substitute of the in-depth exploration of the subject contained in *ML*.

As prerequisites, the reader should have already understood the following statements.

Definition (*ML*: 5.15) The entailment of an efficient cause is called *functional entailment*.

Definition (*ML*: 6.23) A natural system is *closed to efficient causation* if its every efficient cause is (functionally) entailed within the system.

Postulate of Life (*ML*: 11.28) A natural system is an *organism* if and only if it realizes an (M,R)-system.

Theorem (*ML*: 11.29)   *A natural system is an organism if and only if it is closed to efficient causation.*

This sequence of statements is a succinct summary of our answer to the "What Is Life?" question. Life is a phenomenon that sets organisms apart from nonliving systems and dead organisms, and life is manifested through the relations among the processes of metabolism (M) and repair (R). It is through a network of efficient causes that an (M,R)-system models a living system (i.e. 'organism' in its most general sense), so a reductionistic model based strictly on material causation does not qualify.

The defining characteristic of a *living system*, 'closure to efficient causation', anchors on the key concept of *functional entailment*. (Robert Rosen coined the term in Section 5I of his masterwork *Life Itself* [Rosen 1991].) Note that an efficient cause that is entailed is 'function' in both its mathematical sense ('mapping') and its biological sense ('a mode of action by which a thing fulfils its purpose'; *ML*: 0.28). The *imminence* (which I shall define in this book, in Section 7.16) of a mapping f is the collection of all the (functionally) f-entailed entities that can *themselves* entail. *Functional entailment* and *imminence*, the 'local' and 'global' manifestations of the concept, play leading roles in this Opus II of my epic on relational biology; thus the subtitle.

Contents

Exordium
An Introduction to Relational Biology

My 2009 book *ML* has garnered some attention and has engendered/ sustained/renewed interest on the subject of relational biology. The journal *Axiomathes* (the theme of which is 'Where Science Meets Philosophy') dedicated a recent issue (volume 21 number 3, September 2011; [Poli 2011]) to discussing the nuances of *ML*. Entitled 'Essays on *More Than Life Itself*', the special topical issue comprises four essays commenting on *ML* and my responses [Louie 2011] to these comments. The growing interest also led to my being invited to conferences to speak on the subject. This Exordium is a representation of one of these lectures. It is included herein as a review, or a 'refresher of the whys and wherefores', as it were, of concepts considered in detail in *ML*.

E.1 The Interrogative   Science is an activity based on the interrogative: one poses questions about nature and attempts to gain knowledge by answering these questions.

Aristotle contended that one did not really know a 'thing' (which to Aristotle meant a natural system) until one had answered its '*why?*' with its αἴτιον (primary or original 'cause'). In other words, Aristotle's *science* is precisely the subjects for which one seeks the αἰτία to the interrogative '*?*'.

Aristotle's original Greek term αἴτιον (*aition*) was translated into the Latin *causa*, a word which might have been appropriate initially, but which had unfortunately diverged into our contemporary notion of 'cause', as 'that which produces an effect' (more on this shortly). The possible semantic equivocation may be avoided if one understands that Aristotle's original idea had more to do with 'grounds or forms of explanation', so a more appropriate Latin rendering, in retrospect, would probably have been *explanatio*.

E.2 What Is Life?   Biology is the study of life. The ultimate biological question is, then, "What is life?"

This was the question Erwin Schrödinger posed in 1943 and attempted to answer in a series of lectures delivered in Dublin; the corresponding book was published in 1944 [Schrödinger 1944]. With decades of hindsight and further advances in biology, parts of the book may now appear dated. But the originality expressed in this book is not diminished, and the fact that it is still in print is a testimony to its continuing significance.

The Schrödinger question "What is life?" is an abbreviation. A more explicitly posed expansion is

"What distinguishes a living system from a non-living one?"

alternatively,

"What are the defining characteristics of a natural system
                                for us to perceive it as being alive?"

These are epistemological forms of the question.

E.3 The Modelling Relation   *Causality* in the modern sense, the principle that every effect has a cause, is a reflection of the belief that successions of events in the world are governed by definite relations. *Natural Law* posits the existence of these *entailment* relations *and* that this causal order can be *imaged* by implicative order.

A *modelling relation* is a commutative functorial encoding and decoding between two systems. Between a natural system (an object partitioned from the physical universe) N and a formal system (an object in the universe of mathematics) F, the situation may be represented in the following canonical diagram:

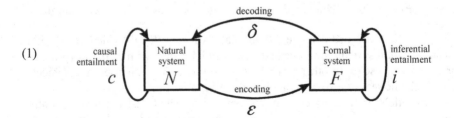

(1)

The encoding ε maps the natural system N and its causal entailment c therein to the formal system F and its internal inferential entailment i; that is,

(2) $\varepsilon : N \to F$ and $\varepsilon : c \to i$.

The decoding δ does the reverse. The entailments satisfy the commutativity condition

(3) $c = \varepsilon \triangleright i \triangleright \delta$.

(Stated graphically, equality (3) says that, in diagram (1), tracing through arrow c is the same as tracing through the three arrows ε, i, and δ in succession.) Thence related, F is a *model* of N, and N is a *realization* of F. In terms of the modelling relation, then, Natural Law is a statement on the existence of causal entailment c and the encodings $\varepsilon : N \to F$ and $\varepsilon : c \to i$.

A formal system may simply be considered as a *set* with additional mathematical structures. So the mathematical statement $\varepsilon : N \to F$, that is, the posited existence for every natural system N a model formal system F, may be stated as the axiom

Everything is a set.

A *mapping* is an inference that assigns to each element of one set a unique element of another set. In elementary mathematics, when the two sets involved are sets of numbers, the inference process is often called a *function*. So 'mapping' may be considered a generalization of the term, when the sets are not necessarily of numbers. (The use of 'mapping' here avoids semantic equivocation and leaves 'function' to its biological meaning.)

Causal entailment in a natural system is a network of interacting processes. The mathematical statement $\varepsilon : c \to i$, that is, the functorial correspondence [*ML*: A.10] between causality c in the natural domain and inference i in the formal domain, may thus be stated as an epistemological principle, the axiom

Every process is a mapping.

Together, the two axioms are the mathematical formulation of Natural Law. These self-evident truths serve to explain "the unreasonable effectiveness of mathematics in the natural sciences".

E.4 Biology Extends Physics   A living system is a material system, so its study shares the material cause with physics and chemistry. Reductionists claim this, therefore, makes biology reducible to 'physics'. *Physics*, in its original meaning of the Greek word φύσις, is simply (the study of) *nature*. So in this sense, it is tautological that everything is reducible to physics. But the hardcore reductionists, unfortunately, take the term 'physics' to pretentiously mean '(the toolbox of) *contemporary* physics'.

Contemporary physics that is the physics of mechanisms reduces biology to an exercise in molecular dynamics. This reductionistic exercise, for example, practised in biochemistry and molecular biology, is useful and has enjoyed popular success and increased our understanding life by parts. But it has become evident that there are incomparably more aspects of natural systems that the physics of mechanisms is *not* equipped to explain.

Biology is a subject concerned with organization of relations. Physicochemical theories are only surrogates of biological theories, because the manners in which the shared matter is organized are fundamentally different.

Hence, the behaviours of the realizations of these mechanistic surrogates are different from those of living systems. This in-kind difference is the impermeable dichotomy between *predicativity* and *impredicativity*. (I shall explicate these two antonyms presently.)

In his 1944 book, Schrödinger wrote:

> "... living matter, while not eluding the 'laws of physics' as established up to date, is likely to involve 'other laws of physics' hitherto unknown, which however, once they have been revealed, will form just as integral a part of science as the former."

There have, of course, been many interpretations of what these 'other laws of physics' might have been. Schrödinger himself likely thought of extensions in thermodynamical terms. It is, however, nothing new in the history of physics that 'other laws of physics' have been added to the repertoire from time to time when 'the toolbox of contemporary physics' became inadequate. The mathematical toolbox of calculus was sufficient for Newtonian mechanics. Tensor geometry had to be recruited for relativity. Operator theory was the appropriate mathematical language of quantum physics. I contend that biology extends physics, and to accordingly expand the toolbox, one needs to enlist *category theory*.

Any question becomes unanswerable if one does not permit oneself a large enough universe to deal with the question. The failure of presumptuous reductionism is that of the inability of a small surrogate universe to exhaust the real one. Equivocations create artefacts. The limits of mechanistic dogma are very examples of the restrictiveness of self-imposed methodologies that fabricate non-existent artificial 'limitations' on science and knowledge. The limitations are due to the nongenericity of the methods and their associated bounded microcosms. One learns something new and fundamental about the universe when it refuses to be exhausted by a posited method.

E.5 Relational Biology The study of biology from the standpoint of 'organization of relations' is a subject called *relational biology*. It was founded by Nicolas Rashevsky (1899–1972) in the 1950s, thence continued and flourished under his student Robert Rosen (1934–1998), my PhD supervisor.

The essence of reductionism in biology is to keep the matter of which an organism is made, and throw away the organization, with the belief that, since physicochemical *structure implies function*, the organization can be effectively reconstituted from the analytic material parts.

Relational biology, on the other hand, keeps the organization and throws away the matter; *function dictates structure*, whence material aspects are entailed.

In terms of the modelling relation, reductionistic biology is physicochemical process seeking models, while relational biology is organization seeking realizations. Stated otherwise, reductionistic biology begins with the material system and relational biology begins with the mathematics. Thus, the principles of relational biology may be considered the operational inverse of (and complementary to) reductionistic ideas. It must be emphasized that both approaches are valuable, each answering questions that the other is not equipped to answer. 'Structure implies function' has beneficial epistemological implications, while 'function dictates structure' better addresses ontological issues.   What renders hardcore reductionism a falsehood is their practitioners' overreaching claim of genericity, their indignant exclusion of other approaches (which they presumptuously consider to be illegitimate), and their self-declared exclusive ownership of objectivity besides. One world is not enough.

In the relational-biological approach, the answer to our "What is life?" question will define an organism as a material system that realizes a certain kind of relational pattern, whatever the particular material basis of that realization may be. For the remainder of this exposition, I shall proceed to answer this question and use the process of reaching this goal to illustrate the methods of relational biology.

E.6 Mapping and Its Relational Diagram   In relational biology, we begin with a formal system, with biology entailed as its realization. So let me begin with a mathematical object, a *mapping* f from set A to set B. It is commonly denoted thus:

(4) $$f:A \rightarrow B.$$

The mapping (4) may alternatively be represented in its category-theoretic notation

(5) $$f \in H(A, B),$$

where $H(A,B)$ denotes a set of mappings from set A to set B and is called a *hom-set*. Essentially, (5) says that $H(A,B)$ is a collection of mappings from set A to set B, and f, being a member of this collection, is one such mapping.

Another way to represent the mapping (4) is its 'element-chasing' version: if $a \in A$, $b \in B$, and the variables are related as $b = f(a)$, then one may use the 'maps to' arrow (note the short vertical line segment at the tail of the arrow) and write

(6) $$f:a \mapsto b.$$

Let me introduce a final representation of the mapping f, its *relational diagram in graph-theoretic form*. It may be drawn as a network with three *nodes* and two *directed edges*, that is, a directed graph (or *digraph* for short):

(7)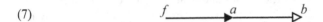

This graph-theoretic representation allows a ready identification of components of a mapping with the four Aristotelian causes that respond to the interrogative "Why mapping?".

The input $a \in A$ is the *material cause*. The output $b \in B$ is the *final cause*. The *hollow-headed arrow* denotes the *flow* from input $a \in A$ to output $b \in B$, whence the final cause of the mapping may be identified also as the hollow-headed arrow that terminates on the output:

(8) $\qquad\qquad\qquad\qquad \longrightarrow\!\!\rhd b$

The *efficient cause* is the *function* of the mapping f as a *processor*; thus, it may be identified as f itself. The *solid-headed arrow* denotes the induction of or constraint upon the flow by the processor f, whence the efficient cause of the mapping may be identified also as the solid-headed arrow that originates from the processor:

(9) $\qquad\qquad\qquad\qquad f \longrightarrow\!\!\blacktriangleright$

The *formal cause* of the mapping is the ordered pair of arrows:

(10) $\qquad\qquad\qquad\qquad \longrightarrow\!\!\blacktriangleright\!\!\longrightarrow\!\!\rhd$

that is, the ordered pair of $\langle$ processor, flow $\rangle$.

E.7 Efficient Cause   Since the efficient cause will turn out to be the crucial *aition* in relational biology, I shall explicate it further.   Aristotle's κινητικός (kinetikos) is rendered into *efficare* in Latin: the efficient cause is "one who puts in motion, that which brings the thing into being, the source of change, that which makes what is made, the 'production rule'".   Note that efficient cause in the Aristotelian sense is simply 'the processor', and the adjective 'efficient' has nothing to do with its common-usage sense that is 'productive with minimum waste or effort'.

The Natural Law axiom "Every process is a mapping." encodes natural processes into mappings; in particular, the encoding identifies an efficient cause of a natural process with the efficient cause of the corresponding mapping. The isomorphic correspondence between the *solid-headed arrow* (9) and the efficient cause of a mapping then completes the linkage in our formalism. Each statement on entailment thus has three analogous formulations, concerning:

i. Causal entailment patterns among efficient causes of natural processes

ii. Inferential entailment paths among efficient causes of mappings

iii. Graphical entailment networks among solid-headed arrows

E.8 Compositions   The relational diagrams of mappings may *interact*: two mappings, with the appropriate domains and codomains, may be connected at different common nodes.

As a first example, consider $g : x \mapsto a$ and $f : a \mapsto b$; thus, *the output of g is the input of f* (the common 'middle' element a). In terms of hom-sets, one has $g \in H(X, A)$ and $f \in H(A, B)$ (where, naturally, $x \in X$, $a \in A$, and $b \in B$); thus, *the codomain of g is the domain of f* (the common 'middle' set A). The relational diagrams of these two mappings connect at the common node a as

(11)

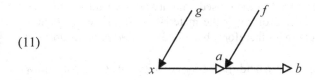

This *sequential composition* of relational diagrams represents the composite mapping $f \circ g \in H(X, B)$ with $f \circ g : x \mapsto b$.

When several mappings are linked by sequential compositions, one has a *sequential chain*:

(12)

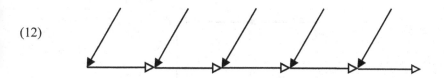

When the first and last mappings in a sequential chain are themselves linked by sequential composition, the chain folds up into a *sequential cycle*:

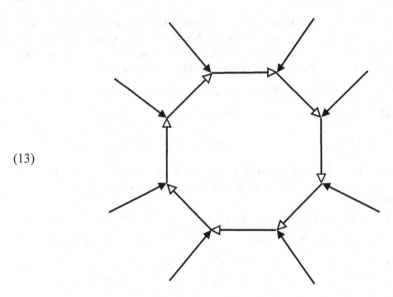

(13)

Note that *within* a sequential cycle, the arrows involved have a consistent direction and are *all hollow-headed* (with solid-headed arrows *peripheral* to the cycle). That is, the compositions involved in the closed path are all sequential, and each final cause has the additional role of being the material cause of the subsequent mapping. A sequential cycle may, therefore, be called a *closed path of material causation.*

Next, consider two mappings g and f with $g:x \mapsto f$ and $f:a \mapsto b$ —now *the output of g is itself the mapping f* .   The hom-sets involved are $g \in H(X, H(A,B))$ and $f \in H(A,B)$: thus, *the codomain of g contains f* . Because of this 'containment', the mapping g may be considered to occupy a higher 'hierarchical level' than the mapping f (and that the hom-set $H(X, H(A,B))$ is at a higher hierarchical level than $H(A,B)$). For these two mappings, one has the *hierarchical composition* of relational diagrams:

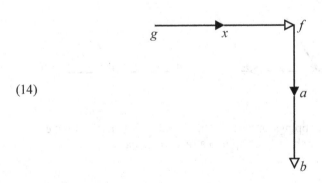

(14)

Since the final cause (i.e. output) of g is the efficient cause of f, the mapping g may be considered an 'efficient cause of efficient cause'. An iteration of efficient causes is inherently hierarchical, in the sense that a lower-level efficient cause is contained within a higher-level efficient cause. In sequential composition, the first mapping g produces something to be operated on, but in hierarchical composition, the first mapping g produces instead an operator itself. Hierarchical composition thus concerns a 'different' mode of entailment, which is given the name of *functional entailment*.

Similar to sequential compositions, hierarchical compositions may form a *hierarchical chain*:

(15)

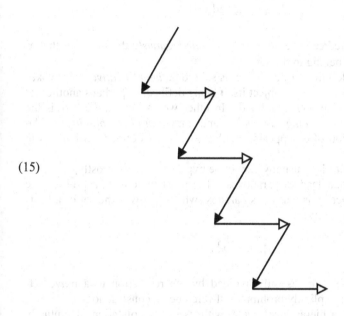

and a *hierarchical cycle*:

(16)

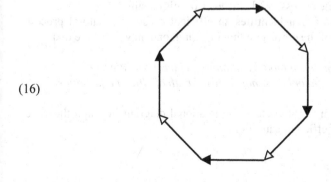

Note that, in contrast to a sequential cycle (13), *solid-headed arrows* (along with hollow-headed arrows) are definitive components of a hierarchical cycle. Efficient causes are relayed; thus, a hierarchical cycle is a *closed path of efficient causation*.

E.9 Impredicativity   In logic, the *predicate* is what is said or asserted about an object. It can take the role as either a property or a relation between entities. Thus, *predicate calculus* is the type of symbolic logic that takes into account the contents (i.e. predicate) of a statement. The defining property $p(x)$ of a subset P in the universe U , as in

$$(17) \qquad\qquad P = \{x \in U : p(x)\},$$

is an example of a predicate, since it *asserts unambiguously* the property that x must have in order to belong to the set P .

   Contrariwise, a definition of an object is said to be *impredicative* if it invokes (mentions or quantifies over) the object itself being defined, or perhaps another set which contains the object being defined. In other words, *impredicativity* is the property of a *self-referencing definition* and may *entail ambiguities*.   An impredicative definition often appears circular, as what is defined participates in its own definition.

   Impredicative definitions usually cannot be bypassed and are mostly harmless. But there are some that lead to paradoxes. The most famous of a problematic impredicative construction is Russell's paradox, which involves the set of all sets that do not contain themselves:

$$(18) \qquad\qquad \{x : x \notin x\} .$$

(This foundational difficulty is only avoided by the restriction to a naive set-theoretic universe that explicitly prohibits self-referencing constructions.)

   It is evident that a hierarchical cycle, with its cyclic collection of mutually entailing efficient causes, is impredicative. In other words, a hierarchical cycle is an *impredicative cycle of inferential entailment*.   A closed path of efficient causation must form a hierarchical cycle of containment: both the hierarchy of containment and the cycle are essential attributes of this closure.

   Through the encoding that identifies an efficient cause of a natural process with the efficient cause of the corresponding mapping, one may conclude that

> *A natural system has a model containing a hierarchical cycle*
> *if and only if it has a closed path of efficient causation.*

Stated otherwise, a hierarchical cycle is the relational diagram in graph-theoretic form of a closed path of efficient causation.

E.10 Nonsimulability   An *algorithm* is a computation procedure that requires in its application a *rigid stepwise mechanical execution of explicitly stated rules*. It is presented as a prescription, consisting of a finite number of instructions. It halts after a finite number of steps. It has no room for ambiguity.

Predicates are algorithmic. Impredicativity is everything that an algorithm is *not*.

A mapping is *simulable* if it is definable by an algorithm. A formal system, an object in the universe of mathematics, may be considered a collection of mappings connected by the system's entailment pattern (i.e. its graph, which may itself be considered a mapping). So by extension, a formal system is *simulable* if its entailment pattern and all of its mappings are simulable. Simulability entails finiteness: that the corresponding Turing machine halts after a *finite* number of steps, that the corresponding algorithmic process is of *finite* length, and that the corresponding program is of *finite* length.

Impredicativity has many consequences. In view of its being the antithesis of things algorithmic, one of these consequences is, therefore, nonsimulability.

Among the entailment networks (12), (13), (15), and (16) that we have considered, the first three, namely, sequential chain, sequential cycle, and hierarchical chain, are simulable, but the last one, hierarchical cycle, is not. The nonsimulability of a hierarchical cycle has been proven using lattice theory. I state this theorem formally as

A formal system that contains a hierarchical cycle is not simulable.

For natural systems, a *deadlock* is a situation wherein competing actions are waiting for one another to finish, and thus none ever does. A set of processes is in a deadlock state when every process in the set is waiting for an event that can be *caused* only by another process in the set. This is a realization, a relational analogue, of impredicativity. In computer science, deadlock refers to a specific condition when two or more processes are each waiting for another to release a resource, or more than two processes are waiting for resources in a circular chain. Implementation of hierarchical cycles (or attempts to execute ambiguous codes in general) will lead a program to either a deadlock or an endless loop. In either case, the program does not terminate. This is practical verification that a hierarchical cycle is *not* simulable.

E.11 Biological Realization: Metabolism-Repair System   Every process is a mapping. The crucial biological process of *metabolism* may, therefore, be represented as a mapping $f : a \mapsto b$ (equivalently, $f \in H(A, B)$); an *enzyme* may be the realization of the efficient cause f, with material input and output metabolites realizations of a and b. Networks of mappings in sequential composition are, then, models of metabolic pathways.

Some biochemical processes produce enzymes as outputs. Such a process may naturally be modelled as a mapping of the form $\Phi : x \mapsto f$ (equivalently, $\Phi \in H(X, H(A, B))$). The morphism Φ may be considered *repair*: its codomain is $H(A,B)$, so it is a mapping that creates new copies of enzymes f, hence a *gene* that 'repairs' (or replenishes) the metabolism process. The repair map Φ and the metabolism map f are thus in hierarchical composition.

A typical eukaryotic cell is compartmentalized into two observably different regions, the cytoplasm and the nucleus. Metabolic activities mainly occur in the cytoplasm, while repair processors (i.e. genes) are contained in the nucleus. Repair in cells generally takes the form of a continual synthesis of basic units of metabolic processor (i.e. enzymes), using as inputs materials provided by the metabolic activities themselves. In particular, the simplest domain of the repair map Φ may be the codomain of metabolism f, the latter's 'output set' B (i.e. $\Phi : b \mapsto f$, $\Phi \in H(B, H(A,B))$), whence metabolism and repair combine into the relational diagram

(19)

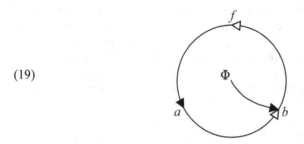

This geometry gives a graphic representation of the metabolism component as the abstract equivalent of 'cytoplasm' and the repair component as the abstract counterpart of 'nucleus'.

What if the repair components themselves need repairing? New mappings representing *replication* (serving to replenish the repair components) may be defined.   A replication map must have as its codomain the hom-set $H(X, H(A,B))$ to which repair mappings Φ belong, so it must be of the form

(20) $\beta : Y \to H\big(X, H(A,B)\big)$

for some set Y (where Y contains ingredients already present in the cell). In the simplest case, when $X = B$, one may choose $Y = H(A,B)$; so (20) becomes

(21) $\beta : H(A,B) \to H\big(B, H(A,B)\big)$.

It turns out that under stringent but not prohibitively strong conditions, the replication mapping β may already be entailed within the components present. There are many ways in which this happens; one natural way is that an isomorphic correspondence may be defined between b and β, whence the mapping (21) may be equivalently represented as

$$(22) \qquad\qquad b: f \mapsto \Phi.$$

The relational diagram of the entailment among the metabolism-repair-replication mappings

$$(23) \qquad\qquad \{\, f : a \mapsto b,\ \Phi : b \mapsto f,\ b : f \mapsto \Phi \,\}$$

is then

(24)

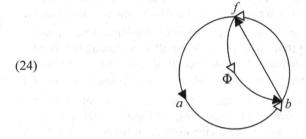

Diagram (24) is the relational diagram in graph-theoretic form of the simplest *metabolism-repair system* (or *(M,R)-system* for short), introduced by Robert Rosen in the late 1950s.

Note that (24) is a hierarchical cycle. The entailment pattern is more evident when the relational diagram is unfolded thus:

(25)

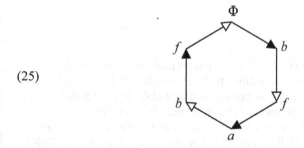

One may also note that there is no 'privileged' position of any of the three mappings involved. They are in cyclic entailment and may be assigned the labels of metabolism, repair, and replication in any cyclic permutation. The all-important feature is that the mappings form a hierarchical cycle; stated otherwise, the simplest (M,R)-system is a hierarchical-cycle model of a cell.

In the specialization of the replication map β from (20) to (21), many simplifying assumptions have been made to create the three-mapping $\{f, \Phi, b\}$ hierarchical cycle. A more sophisticated (M,R)-system model of a cell would contain a large number of metabolism and repair components connected in a complex entailment network, since in a cell there are obviously many more than three interacting processes. (Diagram (24) actually already captures the essence of all (M,R)-systems, and indeed it is possible in principle to reduce every abstract (M,R)-system to this simple form by making the three mappings involved sufficiently complex. One must, nevertheless, not lose sight of the network aspect of (M,R)-systems.)

Metabolism may alternatively be considered an input-output system, with the mapping f representing the transfer function of the 'block', the domain A as the set of inputs, and the codomain B as the set of outputs. Similarly, *repair* may be considered an input-output system, with the mapping Φ representing the transfer function of the block, the domain B as the set of inputs, and the codomain $H(A, B)$ as the set of outputs. With the addition of entailment arrows for environmental inputs and outputs, and the abbreviated representation by the symbols M and R of the components, the relational diagram (19) may be represented as this simple network of one metabolism component and one repair component:

(26)

In general, a metabolism-repair network consists of many metabolism and repair components, with the requisite connections that the outputs of a repair component are observables in the hom-set of its corresponding metabolism component; the metabolism components may be connected among themselves by their inputs and outputs; and repair components must receive at least one input from the outputs of the metabolism components of the network. The following is a sample (M,R)-network (still relatively simple) with six pairs of metabolism-repair components:

(27)

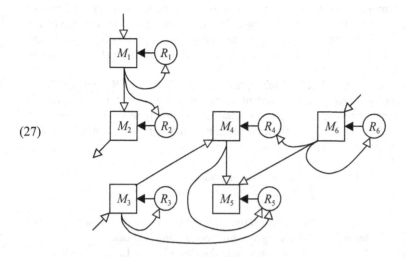

One may easily visualize larger (M,R)-networks with thousands of components.

E.12 Closure to Efficient Causation   Suppose a natural system *contains* a closed path of efficient causation, then *some* of its efficient causes are in cyclic entailment of one another.   Their corresponding mappings must then form a hierarchical cycle. If it so happens that *all* of a natural system's efficient causes entail one another, then it must have a model in which *all* solid-headed arrows are components of hierarchical cycles (e.g. diagram (24) of the simplest (M,R)-system). Having *all* efficient causes entailed within the system is a more stringent requirement than having just *some*, and members of this subset of natural systems are given a special description: *closed to efficient causation*.

> A natural system is *closed to efficient causation* if its every efficient cause is entailed within the system.

The correspondence between an efficient cause and a solid-headed arrow implies:

> A natural system is *closed to efficient causation* if and only if each connected component in its relational diagram has a closed path that contains all the solid-headed arrows.

I mention in passing that "a closed path that contains all the solid-headed arrows" is related to the concept of *traversability* (one continuous trace of the edges in a graph, passing along each edge exactly once) in network topology. Thus, the study of 'closed to efficient causation' can make use of the powerful results from the mathematical theory of topology (in addition to lattice theory and category theory that we have already encountered).

Not all metabolism-repair networks satisfy the stringent requirements for entailment closure. The defining characteristic of an (*M,R*)-*system* that makes it a model of cells is the self-sufficiency in the networks of metabolism and repair components, in the sense that every mapping is entailed within, in short, closure to efficient causation.

The answer to our "What is life?" question according to the Rashevsky-Rosen school of relational biology, in a nutshell, is that an *organism*—the term is used in the sense of an 'autonomous life form', that is, any living system (including, in particular, cells)—admits a certain kind of relational description, that it is 'closed to efficient causation'. Explicitly:

A material system is an organism
if and only if it is closed to efficient causation.

This 'self-sufficiency' in efficient causation is what we implicitly recognize as the one feature that distinguishes a living system from a nonliving one.

In terms of (M,R)-systems, we may state the **Postulate of Life**:

A natural system is an organism
if and only if it realizes an (*M,R*)-*system.*

Thus, an (M,R)-system is the very model of life, and, conversely, life is the very realization of an (M,R)-system.

Prolegomenon
Cardinalis

Not everything that counts can be counted, and not everything that can be counted counts.

— attributed to Albert Einstein

As I did in *ML*, in this book, I assume that the reader is familiar with the basic facts of *naive set theory*, as presented, for example, in Halmos [1960]. In this prologue, however, I shall present some set-theoretic and logical preliminaries; this is more for the clarity of notations (especially for those non-standardized ones) than for the concepts themselves.

Sets

0.1 Definition   If A and B are sets and if every element of A is an element of B, then A is a *subset* of B, denoted

$$(1) \qquad\qquad\qquad A \subset B.$$

Note that this symbolism of containment means *either* $A = B$ (which means the sets A and B have the same elements; *ML*: 0.2: Axiom of Extension) *or* A is a proper subset of B (which means that B contains at least one element that is not in A). Two sets A and B are equal if and only if $A \subset B$ and $B \subset A$ (*ML*: 0.4).

0.2 Definition   If X is a set, the *power set* $\mathcal{P}X$ of X is the family of all subsets of X.

An alternate notation of the power set $\mathcal{P}X$ is 2^X (*cf. ML*: A.3 for the etymology).

0.3 Definition   The *relative complement* of a set A in a set B is the set of elements in B but not in A:

(2) $$B \sim A = \{x \in B : x \notin A\}.$$

When B is the 'universal set' U (of some appropriate universe under study, e.g. the set $\mathbf{N}$ of all natural systems), the set $U \sim A$ is denoted A^c, that is,

(3) $$A^c = \{x \in U : x \notin A\},$$

and is called simply the *complement* of the set A.

0.4 Number Sets   Various sets of numbers are denoted thus:

  i. *Natural numbers* ('positive integers') $\mathbb{N} = \{1, 2, 3, 4, ...\}$
  ii. *Whole numbers* ('nonnegative integers') $\mathbb{N}_0 = \{0\} \cup \mathbb{N} = \{0, 1, 2, 3, 4, ...\}$
  iii. *Integers* $\mathbb{Z} = \{..., -3, -2, -1, 0, 1, 2, 3, ...\}$
  iv. *Rational numbers* ('fractions') $\mathbb{Q} = \left\{ \dfrac{p}{q} : p \in \mathbb{Z}, q \in \mathbb{N} \right\}$
  v. *Real numbers* $\mathbb{R}$
  vi. *Complex numbers* $\mathbb{C}$

The six number sets are related by

(4) $$\mathbb{N} \subset \mathbb{N}_0 \subset \mathbb{Z} \subset \mathbb{Q} \subset \mathbb{R} \subset \mathbb{C}$$

in which all containments are proper.

Equipotence

0.5 Definition   Two sets are *equipotent* (to each other) if there exists a bijective mapping, that is, a one-to-one correspondence, between them (*cf. ML*: 1.8).

Stated otherwise, two sets are equipotent if they are isomorphic in the category **Set** (*cf. ML*: A.6). It is evident that equipotence is an equivalence relation (*ML*: 1.11). The symmetry of the relation also allows the usage 'set A is equipotent to set B', since it implies 'set B is equipotent to set A', whence A and B are equipotent to each other. One also occasionally sees the usage of 'equipollent', or even 'equinumerous', for the same concept.

0.6 Schröder-Bernstein Theorem   *If each of two sets is equipotent to a subset of the other, then the two sets are equipotent.*

Since every set itself is its own subset, the converse of the Schröder-Bernstein Theorem, that if two sets are equipotent then each is equipotent to a subset of the other, is trivially true.

0.7 Law of Trichotomy of Equipotence   *Two sets are either equipotent to each other, or one is equipotent to a subset of the other.*

If two sets are equipotent, then it is easy to see that their power sets are equipotent. But a set is never equipotent to its own power set; this is

0.8 Cantor's Theorem   *Every set is equipotent to a proper subset of its power set, but is not equipotent to the power set itself.*

Cardinality

0.9 Definition   A set is *finite* if it is either empty or equipotent to the set $\{0,1,2,...,n-1\}$ for a natural number n; otherwise it is *infinite*. An infinite set that is equipotent to the set $\mathbb{N}$ of all natural numbers is called *countably infinite*; otherwise the infinite set is *uncountable*. The term *countable* means either finite or countably infinite.

With the formal definition $0 = \varnothing$ and $n = \{0,1,2,...,n-1\}$ for $n \in \mathbb{N}$, a finite set is equivalently 'equipotent to a whole number'. Each finite set X is equipotent to a *unique* whole number $|X| = n \in \mathbb{N}_0$, the 'number of elements of X'. In short, a finite set is a set consisting of a finite number of elements.

The property that each finite set is equipotent to a unique whole number may be extended to infinite sets. The generalized 'number of elements' of a set is called its *cardinality*, and formally one has the

0.10 Property   Every set is equipotent to a unique *cardinal number.*

I will not go into the formal definition of cardinal number (and its related concept ordinal number) here. The interested reader may read Halmos [1960]. The usual partial order $\leq$ of whole numbers may be extended to all cardinal numbers. One uses the same notation $|X| = n$ for the cardinality of the set X, where n may be an 'infinite cardinal' in addition to a whole number. Infinite cardinal numbers are usually denoted by the first letter $\aleph$ (*aleph*) of the Hebrew alphabet. When $|X| = n$, one may simply say 'X has cardinal number n' or 'X has cardinality n'.

When $|X| = n$, a bijective mapping from the cardinal number n (as a set) to the set X is called an *enumeration*, a 'listing of the elements' of X. While 'to enumerate' literally means 'to count out' (i.e. 'to have a number as output'), the domain of an enumeration may be any cardinal number, countable or uncountable. The enumeration map is not uniquely defined by the correspondence $n \leftrightarrow X$, since any permutation of the assignment also serves as a bijection (each different permutation—there being $n!$ of them for finite n—defining its own distinct listing of elements of the set X).

0.11 Theorem

i. *Every set has a cardinal number.*
ii. *Two sets A and B are equipotent if and only if they have the same cardinal number, that is, iff $|A| = |B|$.*
iii. *$|A| \le |B|$ if and only if A is equipotent to a subset of B.*
iv. *$|A| < |B|$ if and only if A is equipotent to a subset of B but B is not equipotent to a subset of A.*

Some trivial properties of finite sets are:

0.12 Corollary

i. *Every finite set has a unique number of elements.*
ii. *Two finite sets are equipotent if and only if they have the same number of elements.*
iii. *If a set is finite, then every one of its subsets is finite.*
iv. *If a finite set X has n elements and a subset $A \subset X$ has k elements, then $k \le n$; further, $k = n$ iff $A = X$.*
v. *If a set is finite, then it is not equipotent to any of its proper subsets.*

Property v, that a finite set is not equipotent to any of its proper subsets, in fact characterizes finite sets. The inverse thus characterizes infinite sets; stated formally:

0.13 Theorem

i. *A set is infinite if and only if it is equipotent to a proper subset of itself.*
ii. *A set is finite if and only if it is not equipotent to any proper subset of itself.*

One also has the following concerning countability:

0.14 Lemma

i. *Every subset of a countable set is countable.*
ii. *Every infinite set has a countably infinite subset.*

0.15 Cardinality of the Power Set   If $|X| = n$, then $|PX| = 2^n$ (for all cardinal numbers n, finite and infinite). The proof is immediate from the fact that PX is equipotent to $\mathbf{Set}(X,2) = 2^X$, the hom-set of all mappings from X to $2 = \{0,1\}$ (*cf. ML*: A.3). One may succinctly write

(5) $$\left| 2^X \right| = 2^{|X|}.$$

0.16 Cantor's Continuum Hypothesis   The cardinality of the set $\mathbb{N}$ of all natural numbers (whence of all countably infinite sets) is denoted by $\aleph_0$. In view of Lemma 0.14.ii, $\aleph_0$ is, then, the least infinite cardinal number. Analogously, the least uncountable cardinal number is usually denoted by $\aleph_1$. In terms of the canonical order relation $\leq$ of cardinal numbers, $\aleph_1$ is the *least* cardinal number strictly following $\aleph_0$.

Cantor's Theorem (Theorem 0.8) dictates that the set $\mathbb{N}$ is equipotent to a subset of its power set $\mathbb{PN}$, but is not equipotent to $\mathbb{PN}$ itself. Whence, it follows from Theorem 0.11.iv and Section 0.15 that $\aleph_0 < \left| \mathbb{PN} \right| = 2^{\aleph_0}$. Since $\aleph_1$ is the least cardinal number larger than $\aleph_0$, one must have

(6) $$\aleph_1 \leq 2^{\aleph_0}.$$

In his famous *continuum hypothesis*, Cantor conjectured that $\aleph_1 = 2^{\aleph_0}$. (The word 'continuum' is used because $2^{\aleph_0}$ is also the cardinal number of the set $\mathbb{R}$ of all real numbers, the 'cardinality of the continuum', usually denoted $\left| \mathbb{R} \right| = c$.)

The consistency of the continuum hypothesis with the usual axioms of set theory has been proven, that is, the equality $\aleph_1 = 2^{\aleph_0}$ is non-contradictory. It has likewise been proven that the continuum hypothesis is independent of the usual axioms of set theory; that is, the inequality $\aleph_1 < 2^{\aleph_0}$ is also non-contradictory.

Indexed Sets

0.17 Indexed Family   Let I and X be sets. A *family of elements in X indexed by I* is a mapping $x : I \rightarrow X$. The domain I is called the *index set* (note the noun adjunct 'index'), an element $i \in I$ is called an *index*, the range $x(I) \subset X$ is called an *indexed set* (note the past participle 'indexed'), and the value $x(i)$ of the mapping x at an index i, written as x_i (whence the element-chasing form of the mapping x may be written as $x : i \mapsto x_i$), is a *term* (or more precisely 'the i th term') of the family. Such a mapping is often denoted

(7) $$\{ x_i \}_{i \in I},$$

and the mapping is also called an *indexed family* (*in* X). Note that notation (7) represents the indexed family, which is a mapping, whence $\left\{ x_i \right\}_{i \in I} \in X^I$), while the notation

(8) $\left\{ x_i : i \in I \right\}$

represents the indexed set (i.e. the range of the indexed family, whence $\left\{ x_i : i \in I \right\} = x(I) \subset X$. The notion of 'the i th term' only makes sense with respect to the indexed family (7) but not the indexed set (8). Occasionally one may simply use $\left\{ x_i \right\}$ for (7) if the index set I is implicitly understood, but this is not good notation (although it is commonly accepted) because of the possible equivocation between the two different entities (7) and (8), essentially the identification of a mapping with its range.

 One may also note that the mapping $x : I \to x(I)$ is surjective (which is simply the statement that a mapping maps *onto* its range), but an indexed family is not required to be injective. Explicitly, it may happen for $i, j \in I$ that $i \neq j$ but $x_i = x_j$; that is, there may be 'duplicated terms'.

0.18 Indexed Family of Sets   An *indexed family of sets* is an indexed family $A : I \to PX$ (of elements in PX), denoted

(9) $\left\{ A_i \right\}_{i \in I} ,$

where each $A_i \subset X$.

0.19 Indexed Partition   An *indexed partition* of a set X is an indexed family of *nonempty* sets $A : I \to PX$ for which the collection of subsets $\left\{ A_i : i \in I \right\}$ forms a *partition* of X (*cf. ML*: 1.16), that is, for each $i \in I$, $A_i \neq \varnothing$, and

(10) $X = \bigcup_{i \in I} A_i$

with

(11) $A_i \cap A_j = \varnothing$ for $i \neq j$.

0.20 Axiom of Choice   If $\left\{A_i\right\}_{i\in I}$ is an indexed family of nonempty sets indexed by a *nonempty* index set I, then there exists an indexed family $\left\{x_i\right\}_{i\in I}$ such that for each $i \in I$, $x_i \in A_i$.

Compare this with the equivalent statement from *ML*: 1.37:   Given a nonempty family $\mathfrak{A}$ of nonempty sets, there is a mapping f with domain $\mathfrak{A}$ such that $f(A) \in A$ for all $A \in \mathfrak{A}$. The correspondence is $\mathfrak{A} = \left\{A_i : i \in I\right\}$. The mapping $f : \mathfrak{A} \to \bigcup_{A \in \mathfrak{A}} A$ (i.e. $f : \left\{A_i : i \in I\right\} \to \bigcup_{i \in I} A_i$) defined by

(12) $$f(A_i) = x_i$$

is called a *choice mapping*. When the index set I is finite, choosing, for each $i \in I$, an x_i from a nonempty set A_i (i.e. defining a choice mapping f) is a simple procedure; not so when I is infinite.   There is no prescription of how infinitely many choices are to be made, and that is why the existence of the choice mapping has to be postulated in an axiom.   It is almost a convention in mathematics that one explicitly acknowledges when a consequence depends on the Axiom of Choice.

Sequences

0.21 Sequence   An indexed family $\left\{x_i\right\}_{i\in I}$ in X with an index set $I = \left\{1,2,...,n\right\}$ (for some natural number n) or $I = \mathbb{N}$ is called a *sequence* (*finite* or *infinite*, respectively) *in* X.

A finite sequence is often written as a list of its terms:

(13) $$\left\{x_i\right\}_{i\in\{1,2,...,n\}} = \left\{x_1, x_2, ..., x_n\right\};$$

so also is an infinite sequence:

(14) $$\left\{x_i\right\}_{i\in\mathbb{N}} = \left\{x_1, x_2, x_3, ...\right\}.$$

Note that in the listing of the elements on the right-hand side of each of (13) and (14), the distinction between indexed family and indexed set is already somewhat blurred (again, this is not good notation but is commonly accepted). A caveat of the blurred listing notation is that in an indexed family, duplicated terms are kept, while in an indexed set, duplicated terms are (almost) always eliminated. Consider, for example, a finite sequence of two vectors v_1 and v_2, with the two vectors identical and nonzero, that is, $v_1 = v_2 \neq 0$. The *sequence* (i.e. indexed family) $\{v_1, v_2\}$ is linearly dependent, but the *set* $\{v_1, v_2\} = \{v_1\}$ is linearly independent, since it consists actually of just one nonzero vector.

0.22 Preorder   The terms of finite and infinite sequences are well-ordered (*ML*: 3.39) by the natural order of integers of their index sets ($\{1, 2, ..., n\}$ and $\mathbb{N}$ respectively).   Thus, one may *truncate*, for example, an infinite sequence $\{x_1, x_2, x_3, ...\}$ *after m terms* (where $m \in \mathbb{N}$) to split off the infinite 'tail end' $\{x_{m+1}, x_{m+2}, x_{m+3}, ...\}$ and obtain the finite sequence $\{x_1, x_2, ..., x_m\}$. One may say, as another example, that a term x_i *precedes* another term x_j if $i < j$ (and that x_j *follows* x_i in the sequence).

It is important to note that the ordering of the terms in a sequence $\{x_i\}$ has to do with the *positions* of the terms, and not the ordering of the elements themselves in the indexed set. This is because the codomain X is not necessarily equipped with an order, and unless it is, a statement such as $x_i \leq x_j$ is meaningless.

There is, however, a way to *define* the binary relation of *precedence* on the *range* $x(I) = \{x_i : i \in I\}$ using the order inherent in $I \subset \mathbb{N}$, by

(15) $x_i \preceq x_j$ in $x(I)$ iff $i \leq j$ in $\mathbb{N}$.

Note that the relation $\preceq$ is defined only on the range $x(I)$ and not on the rest of the codomain $X \sim x(I)$. It is easy to see that the relation of precedence on $x(I)$ is reflexive and transitive (*ML*: 1.10), but not necessarily either antisymmetric or symmetric. A relation that is reflexive and transitive is called a *preorder* (something that is 'not quite' a partial order, *ML*: 1.20, or an equivalence, *ML*: 1.11). A set equipped with a preorder is called a *preordered set* or *proset*. (I shall revisit binary relations, especially these with special properties, in Chapter 3 of this book.)

Each preordered set $\langle S, \preceq \rangle$ is itself a category (*cf. ML*: A.1). This category S has objects the elements of S, and for $a, b \in S$, the hom-set $S(a, b)$ either contains a single S-morphism or is empty, depending on whether $a \preceq b$ or not. Transitivity of $\preceq$ provides for the composition of morphisms, and reflexivity provides the identity morphisms in $S(a, a)$. (I have discussed *poset*, i.e. partially ordered set, as category in *ML*: 1.31, but indeed a *proset* suffices; the antisymmetry is not needed.)

0.23 Monotonic Sequence As a mapping $x : \langle I, \leq \rangle \to \langle x(I), \preceq \rangle$ of prosets, the sequence $\{x_i\}$ preserves the ordering relation by the very definition of $\preceq$ in (15), and is therefore a morphism in the category of prosets and order-preserving mappings. But the order-preserving property of a sequence $\{x_i\}$ may also exist, as a mapping $x : \langle I, \leq \rangle \to \langle x(I), \preceq \rangle$ of prosets, when the codomain X is already equipped with its own preorder $\preceq$ (even when $\preceq$ is not the precedence defined on $x(I) \subset X$; in particular, when $\langle X, \preceq \rangle$ is in fact a *poset*). As the mapping $x : \langle I, \leq \rangle \to \langle X, \preceq \rangle$ of prosets, the sequence $\{x_i\}$ is *isotone* (*cf. ML*: 1.23) if

$$(16) \qquad\qquad i \leq j \text{ in } I \quad \Rightarrow \quad x_i \preceq x_j \text{ in } X.$$

An isotone sequence $\{x_i\}$ is more commonly called *monotonically increasing*, and implication (16) is equivalent to

$$(17) \qquad\qquad x_i \preceq x_{i+1} \text{ for } i \in \mathbb{N}$$

(or for $i \in \{1, 2, ..., n-1\}$ in the case of a finite sequence). (The sequence $\{x_i\}$ is, of course, monotonically increasing with respect to the relation of precedence on $x(I)$.) If the mapping $x : \langle I, \leq \rangle \to \langle X, \preceq \rangle$ is order reversing (i.e. 'antitone'), then

$$(18) \qquad\qquad i \leq j \text{ in } I \quad \Rightarrow \quad x_i \succeq x_j \text{ in } X,$$

which is equivalent to

$$(19) \qquad\qquad x_i \succeq x_{i+1} \text{ for } i \in \mathbb{N}$$

(or, again, for $i \in \{1, 2, ..., n-1\}$ in the case of a finite sequence); such is a *monotonically decreasing* sequence.

If the ordering in (17) is strict, that is, $x_i \prec x_{i+1}$ which means '$x_i \preceq x_{i+1}$ and $x_i \neq x_{i+1}$' (*ML*: 1.22), then the sequence $\{x_i\}$ is *strictly increasing*. Likewise, a strict inequality $x_i \succ x_{i+1}$ ($x_i \succeq x_{i+1}$ and $x_i \neq x_{i+1}$) in (19) defines a *strictly decreasing* sequence. The class of *monotonic sequences* consists of all the increasing and the decreasing sequences.

0.24 Subsequence   There is an important way of obtaining new sequences from a given infinite sequence $\left\{x_i\right\}_{i\in\mathbb{N}} = \left\{x_1, x_2, x_3, ...\right\}$. Let $\left\{n_k\right\}_{k\in\mathbb{N}} = \left\{n_1, n_2, n_3, ...\right\}$ be an infinite sequence in $\mathbb{N}$ such that

(20) $n_{k+1} > n_k$ for $k \in \mathbb{N}$

(i.e. the sequence $n : \langle \mathbb{N}, \leq \rangle \rightarrow \langle \mathbb{N}, \leq \rangle$ is strictly increasing).   The sequence $k \mapsto x_{n_k}$ is called a *subsequence* of the sequence $\left\{x_i\right\}_{i\in\mathbb{N}} = \left\{x_1, x_2, x_3, ...\right\}$ and is denoted

(21) $\left\{x_{n_k}\right\}_{k\in\mathbb{N}} = \left\{x_{n_1}, x_{n_2}, x_{n_3}, ...\right\}.$

One may see that the subsequence $\left\{x_{n_k}\right\}_{k\in\mathbb{N}}$ is simply the (sequential) composition (*ML*: 5.13) of the mapping $k \mapsto n_k$ (in $\mathbb{N}^{\mathbb{N}}$) followed by the mapping $n \mapsto x_n$ (in $X^{\mathbb{N}}$).

One may also readily verify that every sequence $\left\{x_i\right\}$ is a subsequence of itself, and if $\left\{z_i\right\}$ is a subsequence of $\left\{y_i\right\}$ and $\left\{y_i\right\}$ is a subsequence of $\left\{x_i\right\}$, then $\left\{z_i\right\}$ is a subsequence of $\left\{x_i\right\}$.   Stated otherwise, the relation 'is a subsequence of' on the set $X^{\mathbb{N}}$ of all infinite sequences in X is reflexive and transitive; it is, therefore, a preorder (Section 0.22).   Trivially, the relation 'is a subsequence of' is not symmetric, so it is not an equivalence relation; that it is not antisymmetric (whence not a partial order) may be seen in the following example. Let

(22) $\left\{x_i\right\} = \left\{0,1,0,1,0,1,...\right\}$ and $\left\{y_i\right\} = \left\{1,0,1,0,1,0,...\right\}.$

The mapping $n : k \mapsto k+1$, that is, the sequence

(23) $\left\{n_k\right\} = \left\{2,3,4,5,6,7...\right\},$

is such that

(24) $y \circ n = x$ and $x \circ n = y,$

that is,

(25) $\left\{y_{n_k}\right\} = \left\{x_i\right\} = \left\{0,1,0,1,0,1,...\right\}$ and $\left\{x_{n_k}\right\} = \left\{y_i\right\} = \left\{1,0,1,0,1,0,...\right\}.$

So $\{x_i\}$ and $\{y_i\}$ are subsequences of each other, but $\{x_i\} \neq \{y_i\}$.

0.25 Enumerating Sequence   Recall (Definition 0.9) that a nonempty set is finite if it is equipotent to the set $\{0,1,2,...,n-1\} \cong \{1,2,...,n\}$ for a natural number n, and a set is countably infinite if it is equipotent to the set $\mathbb{N}$ of all natural numbers. Equipotence implies that a nonempty finite set X with cardinality $|X| = n$ has a *bijective indexed family* $x:\{1,2,...,n\} \to X$ listing its elements in order and representing it as a finite sequence $\{x_i\}_{i\in\{1,2,...,n\}} = \{x_1,x_2,...,x_n\}$. (This means, in particular, that for $i,j \in I$, if $i \neq j$, then $x_i \neq x_j$.) Similarly, a countably infinite set X (with cardinality $|X| = \aleph_0$) may be represented as an infinite sequence $\{x_i\}_{i\in\mathbb{N}} = \{x_1,x_2,x_3,...\}$ with its corresponding *bijective* indexed family $x:\mathbb{N} \to X$. The bijective indexed family x, a mapping turning a countable set into a sequence, is called an *enumeration* of X (*cf.* Section 0.10).

For both finite and infinite sets, the choice of the enumeration is, as previously mentioned, not unique: any permutation of the assignment also serves as an enumeration (each different permutation defining its own distinct listing of elements and sequential representation of the set).

PART I
Pentateuchus
Becoming Mapping

He had brought a large map representing the sea,
    Without the least vestige of land:
And the crew were much pleased when they found it to be
    A map they could all understand.

"What's the good of Mercator's North Poles and Equators,
    Tropics, Zones, and Meridian Lines?"
So the Bellman would cry: and the crew would reply
    "They are merely conventional signs!

Other maps are such shapes, with their islands and capes!
    But we've got our brave Captain to thank"
(So the crew would protest) "that he's bought *us* the best —
    A perfect and absolute blank!"

      — Lewis Carroll (1876)
      The Hunting of the Snark
      Fit the Second (The Bellman's Speech)

This introductory Part I is an exploration in five chapters of the algebraic theory of set-valued mappings.

My emphasis is on the topics that will be of use to us on our continuing journey in relational biology.   Some theorems will only be stated in this introduction without proofs; their proofs may be found in Chapter 1 of Aubin and Frankowska [1990], Chapter 2 of Berge [1963], or Chapter 1 of Burachik and Iusem [2008]. These are among the very few books that contain the subject of set-valued mappings, and even therein, the algebraic theory is only a prelude that is quickly passed over to concentrate on the analytic and topological aspects.   I should note that the 'forked arrow' notation $F : X \multimap Y$, to be introduced in Section 2.1, for a set-valued mapping F from set X to set Y, is my own.

1
Mapping Origins

He drove out the man; and at the east of the garden of Eden he
placed the cherubim, and a sword flaming and turning to guard
the way to the tree of life.

— *Genesis* 3:24

In Principio: Mappings

1.1 Definition   Given two sets X and Y, one denotes by $X \times Y$ the set of all
ordered pairs of the form (x,y) where $x \in X$ and $y \in Y$. The set $X \times Y$ is
called the *product* (or *Cartesian product*) of the sets X and Y.

The definition of product may be extended to any finite sequence of sets (*cf.*
Sections 0.18 and 0.21) $\left\{ X_i \right\}_{i \in \{1,2,\dots,n\}} = \left\{ X_1, X_2, \dots, X_n \right\}$, of which the product is

the set of all *ordered n-tuples* of the form $(x_1, x_2, \dots, x_n)$ where, for $i = 1, 2, \dots, n$,
$x_i \in X_i$, and may alternatively be denoted

$$(1) \qquad X_1 \times X_2 \times \cdots \times X_n = \prod_{i=1}^{n} X_i = \prod_{i \in (1,2,\dots,n)} X_i$$

(the Cartesian product being the product in the category **Set**; *ML*: A.26).

If either X or Y is empty, then $X \times Y$ is empty. If $X \neq \varnothing$ and $Y \neq \varnothing$, then
there is an element $x \in X$ and an element $y \in Y$, whence $(x,y) \in X \times Y$ and
$X \times Y \neq \varnothing$. These remarks may trivially be extended to a finite sequence of sets
$\left\{ X_i \right\}_{i \in \{1,2,\dots,n\}} = \left\{ X_1, X_2, \dots, X_n \right\}$ — $X_1 \times X_2 \times \cdots \times X_n = \varnothing$ if and only if at least

one $X_i = \varnothing$. For an *infinite* indexed family of sets (i.e. $\left\{ X_i \right\}_{i \in I}$ where the index
set I is infinite), the sufficiency of the previous statement is still trivial: if at least

A.H. Louie, *The Reflection of Life*, IFSR International Series on Systems Science
and Engineering 29, DOI 10.1007/978-1-4614-6928-5_1,
© Springer Science+Business Media New York 2013

one $X_i = \emptyset$, then $\prod_{i \in I} X_i = \emptyset$. The necessity, however, is nontrivial, and the inverse statement is in fact an alternate statement of the

1.2 Axiom of Choice   The product of a nonempty family of nonempty sets is nonempty.

(For a review of necessity versus sufficiency and the logic of conditional statements in general, see the Prolegomenon of *ML*.) Stated otherwise, if $\{X_i\}_{i \in I}$ is a family of nonempty sets indexed by a nonempty set I, then there exists an indexed family $\{x_i\}_{i \in I}$ such that for each $i \in I$, $x_i \in X_i$ (which is the Axiom of Choice stated in 0.20). The 'ordered I-tuple' $(x_i)_{i \in I}$ is an element in the product $\prod_{i \in I} X_i$, whence $\prod_{i \in I} X_i \neq \emptyset$.

1.3 Definition   A *relation* is a set R of ordered pairs; that is, $R \subset X \times Y$ for some sets X and Y.

If $(x, y) \in R$, then one may say that x *is related to* y.

    Equivalently, a relation R is an element of the power set $P(X \times Y)$ (Definition 0.2), that is, $R \in P(X \times Y)$. The collection of *all* relations between two sets X and Y is thus the power set $P(X \times Y)$. The relation $U = X \times Y \in P(X \times Y)$ is the *universal relation*, in which every $x \in X$ is related to every $y \in Y$. The relation $\emptyset \in P(X \times Y)$ is the *empty relation*, in which no $x \in X$ is related to any $y \in Y$. In the partially ordered set $\langle P(X \times Y), \subset \rangle$, U is the greatest element and $\emptyset$ is the least element (*cf. ML*: 1.28). For all relations $R \in P(X \times Y)$, $\emptyset \subset R \subset U$.

1.4 Definition   A *mapping* is a set f of ordered pairs with the property that, if $(x, y) \in f$ and $(x, z) \in f$, then $y = z$.

    Note that the requirement for a subset of $X \times Y$ to qualify as a mapping is in fact quite a stringent one: an element $x \in X$ cannot be related to more than one element of $y \in Y$. Most relations, that is, *generic* members of $P(X \times Y)$, do not have this 'single-valued' property.

1.5 Definition   Let f be a mapping. One defines two sets, the *domain* of f and the *range* of f, respectively, by

(2)
$$\text{dom}(f) = \{x \in X : (x,y) \in f \text{ for some } y \in Y\}$$

and

(3)
$$\text{ran}(f) = \{y \in Y : (x,y) \in f \text{ for some } x \in X\}.$$

Thus $\text{dom}(f) \subset X$ and $\text{ran}(f) \subset Y$, and f is a subset of the product $\text{dom}(f) \times \text{ran}(f)$. If $\text{ran}(f)$ contains exactly one element, then f is called a *constant mapping*.

Various words, such as 'function', 'transformation', and 'operator', are used as synonyms for 'mapping'. The mathematical convention is that these different synonyms are used to denote mappings having special types of sets as domains or ranges. Because these alternate names also have interpretations in biological terms, to avoid semantic equivocation, in this book I shall—unless convention dictates otherwise—use *mapping* (and often *map*) for the mathematical entity.

1.6 Notations   The traditional concept of a mapping is that which assigns to each element of a given set a definite element of another given set; that is, a 'point-to-point' map. I shall now reconcile this with the formal definition given above. Let X and Y be sets, and let $f \subset X \times Y$ be a mapping. This implies $f \subset \text{dom}(f) \times \text{ran}(f) \subset X \times Y$. If one further requires that $\text{dom}(f) = X$ (I shall have more to say about this restriction in Section 1.24 below.), then one says that f is a *mapping of X into Y*, denoted by

(4)
$$f : X \to Y$$

and occasionally (mostly for typographical reasons) by

(5)
$$X \xrightarrow{\ f\ } Y.$$

The collection of all mappings of X into Y is a proper subset of the power set $P(X \times Y)$; this subset is denoted Y^X. Suggestively, one has $Y^X \subset 2^{X \times Y}$.

To each element $x \in X$, by Definition 1.4, there corresponds a unique element $y \in Y$ such that $(x,y) \in f$. Traditionally, y is called the *value of the mapping f at the element x*, and the relation between x and y is denoted by $y = f(x)$ instead of $(x,y) \in f$. Note that the $y = f(x)$ notation is only logically consistent when f is a mapping (i.e. single-valued). For a general relation f, it is possible that $y \neq z$ yet both $(x,y) \in f$ and $(x,z) \in f$; if one were to write $y = f(x)$ and $z = f(x)$ in such a situation, then one would be led to the conclusion that $y = z$: a direct contradiction to $y \neq z$.

With the $y = f(x)$ notation, one has

(6) $$\operatorname{ran}(f) = \{ y \in Y : y = f(x) \text{ for some } x \in X \},$$

which may be further abbreviated to

(7) $$\operatorname{ran}(f) = \{ f(x) : x \in \operatorname{dom}(f) \}.$$

One then also has

(8) $$f = \{ (x, f(x)) : x \in X \}.$$

From this last representation, we observe that when $X \subset \mathbb{R}$ and $Y \subset \mathbb{R}$ (where $\mathbb{R}$ is the set of real numbers), my formal definition of a mapping coincides with that of the 'graph of f' in elementary mathematics.

1.7 Element Chase   Sometimes it is useful to trace the path of an element as it is mapped. If $a \in X$, $b \in Y$, and $b = f(a)$, one uses the 'maps to' arrow (note the short vertical line segment at the tail of the arrow) and writes

(9) $$f : a \mapsto b.$$

One occasionally also uses the 'maps to' arrow to define the mapping f itself:

(10) $$x \mapsto f(x).$$

Mappings of Sets

1.8 Definition   Let f be a mapping of X into Y . If $E \subset X$, the *image* of E under f is defined to be the set $f(E)$ of all elements $f(x) \in Y$ for $x \in E$; that is,

(11) $$f(E) = \{ f(x) : x \in E \} \subset Y.$$

In this notation, $f(X)$ is the range of f . One may also note that, for all $x \in X$,

(12) $f(\{x\}) = \{f(x)\}$.

1.9 Definition   If f is a mapping of X into Y , the set Y is called the *codomain* of f , denoted by $\mathrm{cod}(f)$.

While the domain and range of $f : X \to Y$ are specified by f in $\mathrm{dom}(f) = X$ and $\mathrm{ran}(f) = f(X)$, the codomain is not yet uniquely determined. All that is required so far is that the codomain contains the range as a subset, $Y \supset \mathrm{ran}(f)$. One needs to invoke a category theory axiom that assigns to each mapping f a unique set $Y = \mathrm{cod}(f)$ as its codomain. The axiom is on the mutual exclusiveness of hom-sets in a category $\mathbf{C}$:

(13) $\mathbf{C}(A,B) \cap \mathbf{C}(C,D) = \varnothing$ unless $A = C$ and $B = D$.

Thus each $\mathbf{C}$-morphism f determines a unique pair of $\mathbf{C}$-objects, its domain $A = \mathrm{dom}(f)$ and codomain $B = \mathrm{cod}(f)$, such that $f \in \mathbf{C}(A,B)$. One may consider that associated with a category $\mathbf{C}$ there is a pair of 'mappings' (hence with unique images), dom and cod, that takes $\mathbf{C}$-morphisms to $\mathbf{C}$-objects. (I shall elaborate on this pair of mappings in Sections 6.8 and 6.9.) Alternatively, one may consider a $\mathbf{C}$-morphism as a *triple* (A, B, f) consisting of two $\mathbf{C}$-objects A , B and a $\mathbf{C}$-morphism $f \in \mathbf{C}(A,B)$; equality between triples occurs when they are component-wise equal.

1.10 The Category Set   The category in which the collection of objects is the collection of all sets (in a suitably naive universe of small sets) and the morphisms are mappings is denoted **Set**. Given two sets X and Y , the hom-set $\mathbf{Set}(X,Y)$ is the collection Y^X of all mappings from X to Y .

I often employ non-full subcategories of **Set**, and I use $H(X,Y)$ for appropriate subsets of $Y^X = \mathbf{Set}(X,Y)$ under consideration (e.g. when mappings $f : X \to Y$ represent metabolic functions). These collections of hom-sets $H(X,Y)$ in the subcategory, of course, still have to satisfy the category axioms.

Axiom (13) is interpreted in the category **Set** to yield unique codomains. If a given mapping f from A to B in fact maps A into a proper subset B' of B , then (A, B, f) and (A, B', f) count as different **Set**-morphisms, although as 'mappings' they are the same. For an illustration, consider the mapping $f : \mathbb{R} \to \mathbb{R}$ defined by $f(x) = x^2$ versus the mapping $g : \mathbb{R} \to \{y \in \mathbb{R} : y \geq 0\}$

defined by $g(x) = x^2$. While f and g are the same 'squaring mapping', they are different as **Set**-morphisms, $(\mathbb{R}, \mathbb{R}, f) \neq (\mathbb{R}, \{y \in \mathbb{R} : y \geq 0\}, g)$.

1.11 Surjection   The range $f(X) = \mathrm{ran}(f)$ is a subset of the codomain $Y = \mathrm{cod}(f)$, but they need not be equal. When they are, that is, when $f(X) = Y$, one says that f is a *mapping of X onto Y* and that $f : X \rightarrow Y$ is *surjective* (or is a *surjection*). Note that every mapping maps onto its range.

1.12 Definition   If $E \subset Y$, $f^{-1}(E)$ denotes the set of all $x \in X$ that f maps into E:

(14) $$f^{-1}(E) = \{x : f(x) \in E\} \subset X$$

and is called the *inverse image* of E under f.

Note that $f^{-1}(Y) = X$, even though $\mathrm{ran}(f) = f(X)$ may be a proper subset of Y. If $y \in Y$, $f^{-1}(\{y\})$ is abbreviated to $f^{-1}(y)$, whence

(15) $$f^{-1}(y) = \{x \in X : f(x) = y\}.$$

1.13 Injection   Note that $f^{-1}(y)$ may be the empty set or may contain more than one element. If, for each $y \in Y$, $f^{-1}(y)$ consists of at most one element of X, then f is said to be an *injective* (also *one-to-one* or *1-1*) *mapping of X into Y*. Other commonly used names are '$f : X \rightarrow Y$ is an *injection*' and '$f : X \rightarrow Y$ is an *embedding*'. This may also be expressed as follows: f is a one-to-one mapping of X into Y provided $f(x_1) \neq f(x_2)$ whenever $x_1, x_2 \in X$ and $x_1 \neq x_2$.

For $A \subset X$, the embedding $i : A \rightarrow X$ defined by $i(x) = x$ for all $x \in A$ is called the *inclusion map* (of A in X). The inclusion map of X in X is called the *identity map* on X, denoted 1_X.

1.14 Lemma

i. $f : X \rightarrow Y$ is injective iff for every $y \in \mathrm{ran}(f)$, $f^{-1}(y)$ is a singleton set in X.

ii. $f : X \rightarrow Y$ is surjective iff for every nonempty subset $E \subset Y$, $f^{-1}(E)$ is a nonempty subset of X.

1.15 Inverse Mapping   In view of the equivalence in Lemma 1.14.i, when $f : X \to Y$ is injective, it defines an *inverse mapping* $f^{-1} : \mathrm{ran}\,(f) \to X$ (with the mild notational equivocation of each singleton set $f^{-1}(y)$ with the element it contains).   Indeed, as a mapping, f^{-1} is necessarily a *one-to-one* mapping of $\mathrm{ran}\,(f)$ *onto* $X = \mathrm{dom}\,(f)$.

A mapping and its inverse (when it exists) compose to identity mappings; thus,

$$(16) \qquad\qquad f^{-1} \circ f = 1_X \quad \text{but} \quad f \circ f^{-1} = 1_{f(X)}$$

$$\text{(and not necessarily } f \circ f^{-1} = 1_Y \text{).}$$

One also has the following simple

1.16 Lemma   *Let $f : X \to Y$ and $g : Y \to X$ be mappings. If $g \circ f = 1_X$ then f is injective and g is surjective.*

1.17 Bijection   If a mapping $f : X \to Y$ is both one-to-one and onto, that is, both injective and surjective, then f is called *bijective* (or is a *bijection*) and that the mapping f establishes a *one-to-one correspondence* between the sets X and Y.

1.18 The Power Set Functor   The *power set functor* $\mathsf{P} : \mathbf{Set} \to \mathbf{Set}$ assigns to each set X its power set $\mathsf{P}X$ (Definition 0.2) and assigns to each mapping $f : X \to Y$ the mapping

$$(17) \qquad\qquad\qquad \mathsf{P}f : \mathsf{P}X \to \mathsf{P}Y$$

that sends each $A \subset X$ to its image $f(A) \subset Y$.   One readily verifies that this definition satisfies the functorial requirements $\mathsf{P}(g \circ f) = \mathsf{P}(g) \circ \mathsf{P}(f)$ (the mapping that sends a subset A of the domain of f to the subset $g(f(A))$ of the codomain of g) and $\mathsf{P}1_X = 1_{\mathsf{P}X}$ (the identity morphism gets sent to the identity morphism), so P is a covariant functor from **Set** to **Set**.

Dually, the *contravariant power set functor* $\overline{\mathsf{P}} : \mathbf{Set} \to \mathbf{Set}$ assigns to each set X its power set $\mathsf{P}X$ and to each mapping $f : X \to Y$ the mapping

$$(18) \qquad\qquad\qquad \overline{\mathsf{P}}f : \mathsf{P}Y \to \mathsf{P}X$$

that sends each $B \subset Y$ to its inverse image $f^{-1}(B) \subset X$.

Thus a 'point-to-point' mapping $f : X \to Y$ naturally defines two 'point-to-point' mappings, $Pf : PX \to PY$ and $\overline{P}f : PY \to PX$, for which a 'point' is an element of a power set, hence a set. Alternatively (with mild notational equivocation), the 'image map' Pf may be considered a 'set-to-set' mapping f from X to Y, sending subsets of X to subsets of Y, while the 'inverse image map' $\overline{P}f$ may be considered a 'set-to-set' mapping f^{-1} from Y to X, sending subsets of Y to subsets of X.

Note that the traffic $f \mapsto Pf$ (or $f \mapsto \overline{P}f$) from a 'point-to-point' mapping to a 'set-to-set' mapping only goes one way. For a given mapping $g : PX \to PY$ of power sets, there is in general no mapping $f : X \to Y$ for which $g = Pf$ (or $f : Y \to X$ for which $g = \overline{P}f$). This is because, in the covariant case (the argument for the contravariant case being similar), for $x \in X$, one would have to have $f(x) = g(\{x\})$. But there is no guarantee that $g(\{x\}) \subset Y$ is a singleton set, which is what is required for f to be a mapping.

Some properties of the 'set-to-set' mapping $Pf : PX \to PY$ are listed in the following theorem:

1.19 Theorem   *Let $f : X \to Y$ and $A, B \subset X$. Then:*

i. $f(A) \neq \varnothing$ *iff* $A \neq \varnothing$.

ii. $A \subset B \implies f(A) \subset f(B)$.

iii. $f(A \cup B) = f(A) \cup f(B)$.

iv. $f(A \cap B) \subset f(A) \cap f(B)$.

v. $f(B \sim A) \supset f(B) \sim f(A)$.

The simple example of $f(1) = f(2) = 0$, $A = \{1\}$, and $B = \{2\}$ (whence $f(A) = f(B) = \{0\}$) shows why the converse of property ii is false and why properties iv and v are not equalities.

One sees from property iv that the mapping $Pf : PX \to PY$ does not preserve the Boolean algebraic structure of power sets: one does not have $f(A \cap B) = f(A) \cap f(B)$. Note property v implies that

(19) $f(X \sim A) \supset f(X) \sim f(A) = \operatorname{ran}(f) \sim f(A)$,

but in general there is no inclusion relation between the sets $f(X \sim A)$ and $Y \sim f(A)$ (since $Y \sim f(X)$ may be nonempty; i.e. f may not be surjective).

On the other hand, the dual 'set-to-set' mapping $\overline{P}f : PY \to PX$ has the following properties:

1.20 Theorem   Let $f : X \to Y$ and $A, B \subset Y$. Then:

i. $A = \emptyset \Rightarrow f^{-1}(A) = \emptyset$.

ii. $A \subset B \Rightarrow f^{-1}(A) \subset f^{-1}(B)$.

iii. $f^{-1}(A \cup B) = f^{-1}(A) \cup f^{-1}(B)$.

iv. $f^{-1}(A \cap B) = f^{-1}(A) \cap f^{-1}(B)$.

v. $f^{-1}(B \sim A) = f^{-1}(B) \sim f^{-1}(A)$.

Going from Theorem 1.19 to Theorem 1.20, one sees that now in properties iv and v, set inclusion has been replaced by equality. Recall (Section 1.15) that the inverse mapping f^{-1} is necessarily a *bijective* mapping of $\operatorname{ran}(f)$ onto $X = \operatorname{dom}(f)$. These 'improvements' result as a consequence. Theorem 1.20 says that $\overline{P}f : PY \to PX$ is in fact a Boolean algebra homomorphism. Note also that $f^{-1}(\emptyset) = \emptyset$ and $f^{-1}(Y) = X$; thus the least and greatest elements are preserved by the inverse mapping f^{-1}.

The converse of property 1.20.i is not true (in contrast to property 1.19.i). This is again because f itself may not be surjective: if $A \subset Y \sim f(X)$ is a nonempty subset, then one still has $f^{-1}(A) = \emptyset$. But $f^{-1} : \operatorname{ran}(f) \to X$ is surjective, so $f^{-1}(Y) = X$; property 1.20.v then also says that

(20) $$f^{-1}(Y \sim A) = X \sim f^{-1}(A)$$

(contrast this with (19) and its subsequent discussion).

The following properties of the composites of f and f^{-1} may also be readily verified:

1.21 Theorem   Let $f : X \to Y$, $A \subset X$, and $B \subset Y$. Then:

i. $A \subset f^{-1}(f(A))$.

ii. $B \supset f(f^{-1}(B))$.

iii. $f(A \cap f^{-1}(B)) = f(A) \cap B$.

The inclusion relations in properties 1.19.iv and v and properties 1.21.i and ii become equality under special conditions:

1.22 Theorem   *Let* $f : X \to Y$. *The following are equivalent:*

i.   *f is injective.*
ii. *For all* $A \subset X$, $A = f^{-1}(f(A))$.
iii. *For all* $A, B \subset X$, $f(A \cap B) = f(A) \cap f(B)$.
iv. *For all* $A, B \subset X$, $f(B \sim A) = f(B) \sim f(A)$.
v.   *For all* $A \subset X$, $f(X \sim A) \subset Y \sim f(A)$.

1.23 Theorem   *Let* $f : X \to Y$. *The following are equivalent:*

i.   *f is surjective.*
ii. *For all* $B \subset Y$, $B = f(f^{-1}(B))$.
iii. *For all* $A \subset X$, $Y \sim f(A) \subset f(X \sim A)$.

What Is a Mapping?

1.24 Hardy's Idea of a Mapping   A   mapping   (i.e.   'function'   in   the mathematical sense) $y = f(x)$ often possesses three characteristics:

i. y is determined for *every value* of x.
ii. y is determined *uniquely* for each value of x; that is, to each value of x corresponds *one and only one* value of y.
iii. The relation f between x and y is expressed by means of an *analytic formula*, from which the value of y corresponding to a given x may be calculated by direct substitution of the latter.

G. H. Hardy, in his seminal textbook *A Course of Pure Mathematics* [10th edition, 1952], dismissed each of these characteristics as "by no means essential to a function". That characteristic iii is not essential is evident: not all functional correspondences are given by neat formulae such as $y = 3x^2 + x - 2$ and $y = a\sin(bx + c)$. Indeed, the existence of an "analytic formula" depends on the set of "elementary functions" one has in one's toolbox, which is expanded by necessity with the augmentation, when circumstances warrant, of "special functions": consider, for example, $y = \log(x)$, $y = \cosh(x)$, $y = \Gamma(x)$, and $y = Li(x)$. But toolbox collections are finite. From another viewpoint, the negation of iii alludes to the fact that not all mappings are computable.

In all the editions of his book (from first edition (1908) to the final tenth edition (1952), which is still being reprinted and available), Hardy maintained:

> "All that is essential is that there should be some relation
> between x and y such that to some values of x at any rate
> correspond values of y."

Note the quantifier in "*some* values of x" and the plurality of "correspond *values* of y".

That characteristic i is not essential is inherent in the definition of a relation R as any subset of the Cartesian product $X \times Y$. There is no requirement that for each $x \in X$, there has to be a $y \in Y$ such that $(x, y) \in R$. This condition may be passed on to mappings, hence the negation of i: $y = f(x)$ (i.e. $(x, y) \in f$) may only hold for "*some* values". For example, one may consider $f(x) = \sqrt{x}$ as a mapping from $\mathbb{R}$ to $\mathbb{R}$, although $f(x)$ is not determined for $x < 0$, whence $\mathrm{dom}(f) = \{x \in \mathbb{R} : x \geq 0\}$. Since $\mathrm{dom}(f) \subset X$, the issue may be bypassed by restricting f to $X' = \mathrm{dom}(f)$, and considering $f \subset X' \times Y$ instead of $f \subset X \times Y$, then y is determined for *every value* of $x \in X'$. This is commonly practised; with the $f(x) = \sqrt{x}$ example, the mapping is more properly considered as from $X' = \{x \in \mathbb{R} : x \geq 0\}$ to $\mathbb{R}$. Indeed, in the notation $f : X \to Y$ (Section 1.6), the convention is that one implicitly takes $\mathrm{dom}(f) = X$ (unless otherwise stated).

The single-valued requirement of characteristic ii is now standard, universally accepted as an integral part of the definition of a mapping (*cf.* Definition 1.4 above). As I remarked in Section 1.6 above, the notation $y = f(x)$ (attributed to Leonhard Euler) only makes logical sense when $f(x) \in Y$ is uniquely determined. In this context of a mapping being single-valued by definition, the term 'multi-valued mapping' is therefore a misnomer; a mapping has to be single valued to be called 'well defined'. But as Hardy declared as late as 1952, it is at times useful to relax characteristic ii to include "*values* of y".

1.25 Well-Posed Problem

Jacques Hadamard stated that mathematical models of physical phenomena should have the properties that:

i. A solution exists.
ii. The solution is unique.
iii. The solution depends continuously on the data, in some reasonable topology.

The formulation of such a model is termed a *well-posed problem* (and an *ill-posed problem* otherwise). Hadamard's well-posed problem is often used as an explanation of why mappings are defined thus, especially their unique-value requirement: compare Hadamard's three properties with the three characteristics in the previous section.

1.26 Examples of Multi-valued Mappings   There   are,   however,   many
situations in which existence fails, when no output is associated with an input, and
in which uniqueness fails, when more than one output are associated with an input.
When a mapping is not surjective, its inverse is not defined on its codomain: for
$f : X \rightarrow Y$ and $y \in Y \sim \text{ran}(f)$, $f^{-1}(y)$ is not defined (at least not by the role of
f^{-1} as an inversion of f). When a mapping is not injective, its inverse is not
single-valued: for $f : X \rightarrow Y$ and $y \in \text{ran}(f)$, $f^{-1}(y)$ may contain more than
one element. Thus the 'inverse' f^{-1} of a mapping f is not necessarily itself a
mapping. Stated otherwise, the 'inverse' of a **Set**-morphism is not necessarily a
Set-morphism.

As a simple example, consider the inverse of the square mapping $y = x^2$ from
$\mathbb{R}$ to $\mathbb{R}$, that is, real solutions to the equation $y^2 = x$. If x is a negative real
number, there are no real solutions for y. (This is, of course, famously the
genesis of complex numbers.) If x is a positive real number, this equation defines
two values of y corresponding to x, namely, $y = \pm\sqrt{x}$. Indeed, the 'square-root
mapping' is not a mapping, unless one follows the convention that the symbol
$\sqrt{x}$ is defined to mean the *positive square root* of a positive real number x
(whence $-\sqrt{x}$ is the negative square root). The proper 'square-root mapping' is
thus the 'double-valued'

(21) $$x \mapsto \left\{ \sqrt{x}, -\sqrt{x} \right\}.$$

In general, for a complex number $z = re^{i\theta}$ (where $-\pi < \theta \leq \pi$), there are n
(distinct when $z \neq 0$) n th roots of z, given by

(22) $$r^{\frac{1}{n}} e^{i\frac{1}{n}(\theta + 2k\pi)}, \qquad k = 0, 1, ..., n-1.$$

The proper ' n th-root mapping' from $\mathbb{C}$ to $\mathbb{C}$ is thus the multi-valued

(23) $$z \mapsto \left\{ r^{\frac{1}{n}} e^{i\frac{1}{n}\theta}, r^{\frac{1}{n}} e^{i\frac{1}{n}(\theta + 2\pi)}, ..., r^{\frac{1}{n}} e^{i\frac{1}{n}(\theta + 2(n-1)\pi)} \right\}.$$

In complex analysis, there are many situations in which 'multi-valued
mappings' arise, and most of them stem from the *ambiguity* of the argument
$\theta = \arg z$ of a complex number

(24) $$z = re^{i\theta},$$

since θ plus any multiple of 2π may be substituted for θ in (24). The
'mapping' arg is thus not well defined:

(25) $$\arg : z \mapsto \{\theta + 2k\pi : k \in \mathbb{Z}\}.$$

In order to have a (single-valued) mapping, one restricts $\arg : \mathbb{C} \to \mathbb{R}$ to one branch (its *principal branch*), and one defines

(26) $$-\pi < \arg z \le \pi,$$

that is, as the mapping $\arg : \mathbb{C} \to (-\pi, \pi]$.

It is often useful to have domains and codomains of complex mappings as open sets; thus one may even restrict further. For example, the complex logarithm of a complex number $z = re^{i\theta}$ is the multi-valued mapping

(27) $$\log : z \mapsto \{\log r + i(\theta + 2k\pi) : k \in \mathbb{Z}\}.$$

Its principal branch is restricted to the domain $\mathbb{C} \sim \{s : s \le 0\}$ (i.e. the complex plane with a 'slit' along the negative real axis) and defined as

(28) $$\log z = \log |z| + i \arg z$$

(where $-\pi < \arg z < \pi$), whence $\log : \mathbb{C} \sim \{s : s \le 0\} \to \{w \in \mathbb{C} : \operatorname{Re} w > 0\}$.

A more sophisticated treatment replaces complex 'multi-valued mappings' with mappings with Riemann surfaces as domains, but I shall not digress thence.

2
From Points to Sets

> He made loops of blue on the edge of the outermost curtain of
> the first set; likewise he made them on the edge of the outermost
> curtain of the second set; he made fifty loops on the one curtain,
> and he made fifty loops on the edge of the curtain that was in the
> second set; the loops were opposite one another.

> — *Exodus* 36:11–12

Congregatio: Set-Valued Analysis

2.1 Set-Valued Mapping   From the forms of the 'point-to-set mappings'
$F : \bullet \mapsto \{\cdots\}$ in Section 1.26 (*cf.* (21), (23), (25), and (27) therein), one may
naturally proceed to define a set-valued mapping thus:

Definition A   A *set-valued mapping* from set X to set Y is a relation
$F \subset X \times Y$ (Definition 1.3). It may be denoted

$$(1) \qquad\qquad F : X \multimap Y,$$

such that for each $x \in X$,

$$(2) \qquad\qquad F(x) = \{ y \in Y : (x, y) \in F \} \subset Y.$$

Note the *point-to-set* nature of a set-valued mapping (as opposed to 'point-to-point' for a standard mapping; *cf.* Section 1.6). This relaxation of characteristic 1.24.ii thus includes, when $F(x)$ contains more than one element, Hardy's allowance of mappings in which to a point may plurally "correspond *values* of y ". Note, also, the possibility that for some $x \in X$, it may happen that $F(x) = \varnothing$. This relaxation of characteristic 1.24.i thus includes Hardy's allowance of mappings in which values may correspond to only "*some* values of x ".

A.H. Louie, *The Reflection of Life*, IFSR International Series on Systems Science
and Engineering 29, DOI 10.1007/978-1-4614-6928-5_2,
© Springer Science+Business Media New York 2013

Note the special 'forked arrow' $-\subset$ that I have chosen to denote set-valued mappings, in distinction from $\to$ for a standard (single-valued) mapping. In this chapter when I introduce the concept of set-valued mapping and its properties, I shall also use capital letters to denote set-valued mappings, e.g., $F : X -\subset Y$, while use lowercase letters to denote standard mappings, e.g., $f : X \to Y$. This F–versus–f distinction may not, however, necessarily continue in later chapters, but the two different arrows will remain as the characterizing form.

In a set-valued mapping's element-chasing form, one may write

(3) $$F : x \mapsto F(x).$$

The 'source' of F is still a *point* $x \in X$, but now the *value* of the mapping F at the element x is a *set* $F(x) \subset Y$. The source (material cause) and the value (final cause) of a set-valued mapping are thus different in kind from each other, they belonging to different hierarchical levels ('point' versus 'set'). (For a review of the identification of Aristotle's four causes with components of a mapping, see *ML*: Chapter 5.)

A standard (single-valued) mapping (as defined in 1.4) $f : X \to Y$ may be considered a very specialized set-valued mapping $F : X -\subset Y$ such that, for each $x \in X$, the value

(4) $$F(x) = \{f(x)\}$$

is a singleton set. Indeed, one can make the formal definition: a set-valued mapping $F : X -\subset Y$ is called *single-valued* if for each $x \in X$, $F(x)$ is a singleton set. A 'single-valued set-valued mapping' $F : X -\subset Y$ therefore defines a 'standard' mapping $f : X \to Y$ by $f : x \mapsto$ the single element in $F(x)$. Thus 'single-valued set-valued mapping' and 'mapping' are equivalent terms.

Since a set-valued mapping $F : X -\subset Y$ takes its values in the family of subsets of Y (i.e., the power set $\mathsf{P}Y$ of Y), one may alternatively consider

Definition B   A *set-valued mapping* from set X to set Y is a (single-valued) mapping $F : X \to \mathsf{P}Y$.

In algebraic terms, the two definitions are equivalent. In topological terms (*cf.* Hadamard's property iii in 1.25), however, because of the complicated power-set topology of $\mathsf{P}Y$ induced by the topology of Y, it is often advantageous to use Definition A.

2.2 Definition   Let $F : X -\subset Y$ be a set-valued mapping. The *graph* of F is defined as F in its relational form; i.e.,

(5) $F = \{(x, y) \in X \times Y : y \in F(x)\} = \{(x, y) \in X \times Y : (x, y) \in F \} \subset X \times Y.$

(Compare this with the 'graph of f' in Section 1.6.)

2.3 Domain   The *domain* of the set-valued mapping $F : X \multimap Y$ is the set X, denoted by $\mathrm{dom}(F)$.

The word 'domain' is from the Latin *domus*, 'house, home'. Thus the domain of a mapping is the set of values for which the mapping 'feels at home' (in the idyllic and idealistic sense of the set of values that 'do not cause the mapping any trouble'). In addition, the related Latin word *dominus* means 'lord, master' literally 'one who rules the home', or 'one who owns the domain'. Thus the domain of a mapping is the set of values that the mapping 'owns' or 'has control of'.

There is a subtle difference in the definitions of 'domain' of a set-valued mapping and a (single-valued) mapping, as respectively given in 2.3 and 1.5. When a mapping is considered as a relation $f \subset X \times Y$, one has $\mathrm{dom}(f) \subset X$. But, as I mentioned in Section 1.24, in the notation $f : X \to Y$ for a standard mapping, the convention is that one implicitly takes $\mathrm{dom}(f) = X$ (whence for every $x \in X$, $f(x)$ is defined and it is a single element in Y). Contrariwise, for a set-valued mapping $F : X \multimap Y$, F is still defined at those $x \in X$ for which $F(x) = \varnothing$. One has $\mathrm{dom}(F) = X$ in both interpretations of F, as the relation $F \subset X \times Y$ and as the point-to-set mapping $F : x \mapsto F(x)$ from X to $\mathrm{P}Y$.

2.4 Definition   The projections of the graph of F onto its first and second components are, respectively, the *corange* and the *range* of F,

(6) $\mathrm{cor}(F) = \{x \in X : F(x) \neq \varnothing\},$

(7) $\mathrm{ran}(F) = \{y \in Y : y \in F(x) \text{ for some } x \in X\}.$

Thus $\mathrm{cor}(F) \subset X$ and $\mathrm{ran}(F) \subset Y$, and both inclusions may be proper.

$X \sim \mathrm{cor}(F) = \mathrm{dom}(F) \sim \mathrm{cor}(F)$ is the subset of X that contains all those $x \in X$ at which $F(x) = \varnothing$. Note that some authors, however, define the domain of F as $\{x \in X : F(x) \neq \varnothing\}$ instead of X itself. But there are category-theoretic advantages in allowing $F(x) = \varnothing$ for $x \in \mathrm{dom}(F)$. (I shall return to this point when I presently introduce the category **Rel** of sets and relations.) The range of F may also be expressed as

$$(8) \qquad \mathrm{ran}(F) = \bigcup_{x \in X} F(x) \subset Y.$$

F (as a relation in $X \times Y$) is thus a subset of the product $\mathrm{cor}(F) \times \mathrm{ran}(F)$. $x \in \mathrm{cor}(F)$ means there exists $y \in \mathrm{ran}(F)$ such that $(x, y) \in F$; dually, $y \in \mathrm{ran}(F)$ means there exists $x \in \mathrm{cor}(F)$ such that $(x, y) \in F$.

If there exists a subset C of Y such that $F(x) = C$ for all $x \in X$, then F is called a *constant set-valued mapping*. As a relation in $X \times Y$, F is the subset $X \times C$. The constant mapping $f : x \mapsto c$ (where $c \in Y$) thus defines the constant set-valued mapping $F : x \mapsto \{c\}$. The universal relation $U = X \times Y$ from X to Y (*cf.* Section 1.3) is the constant set-valued mapping $U : X \multimap Y$ that sends everything to the set Y, i.e., such that $F(x) = Y$ for all $x \in X$.

2.5 The Constant Empty-Set-Valued Mapping   The   constant   set-valued mapping $F : X \multimap Y$ that sends everything to the empty set, i.e., such that

$$(9) \qquad F(x) = \varnothing \quad \text{for all } x \in X,$$

has

$$(10) \qquad \mathrm{cor}(F) = \{x \in X : F(x) \neq \varnothing\} = \varnothing,$$

$$(11) \qquad \mathrm{ran}(F) = \varnothing,$$

and

$$(12) \qquad \{x \in X : F(x) = \varnothing\} = X \sim \mathrm{cor}(F) = X.$$

As a relation in $X \times Y$, F is thus the 'empty relation' $\varnothing$ (*cf.* Section 1.3).

Note that the 'empty relation' $\varnothing$ is a legitimate set-valued mapping from set X to set Y, for all sets X and Y. This is in contrast to standard mappings, when the 'empty mapping' $\varnothing : X \to Y$ is only a mapping when $X = \varnothing$. Recall (ML: A.4) that by convention $Y^\varnothing = \{\varnothing\}$; thus the 'empty mapping' $\varnothing$ is the only mapping from the empty set to any set Y. If $X \neq \varnothing$, however, then $f(X) \neq \varnothing$ for any mapping f with $\mathrm{dom}(f) = X$, whence $\mathrm{ran}(f) \neq \varnothing$; so one has $\varnothing^X = \varnothing$ whence $\varnothing \notin \varnothing^X$.

It is interesting to note that for any two sets X and Y, whatever their nature, the constant empty-set-valued mapping $\varnothing : X \multimap Y$ is the same one. There is only one constant empty-set-valued mapping because there is only one empty set. Suppose $\varnothing_1$ and $\varnothing_2$ are two empty sets. Then $x \in \varnothing_1 \Rightarrow x \in \varnothing_2$, since there is no $x \in \varnothing_1$ to contradict this statement; thus $\varnothing_1 \subset \varnothing_2$. Likewise, $\varnothing_2 \subset \varnothing_1$. Therefore, $\varnothing_1 = \varnothing_2$.

The map that is a 'perfect and absolute blank' of Lewis Carroll's Bellman is an example of a constant empty-set-valued mapping (indeed, a manifestation of *the* empty set) $\varnothing$. As a material system, a blank sheet of paper is, of course, *structurally* nonempty, but, as a map, it *functions* as the empty set.

2.6 Definition   For a set-valued mapping $F : X \multimap Y$, the set Y is called the *codomain* of F, denoted by $\mathrm{cod}(F)$.

Thus one has the dual relations

(13) $$\mathrm{ran}(F) \subset \mathrm{cod}(F) = Y, \quad \mathrm{cor}(F) \subset \mathrm{dom}(F) = X.$$

2.7 Definition   A set-valued mapping $F : X \multimap Y$ is:

i. *Surjective* if

(14) $$\mathrm{ran}(F) = \mathrm{cod}(F) = Y$$

ii. *Semi-single-valued* if

(15) $$F(x_1) \cap F(x_2) \neq \varnothing \;\Rightarrow\; F(x_1) = F(x_2)$$

iii. *Injective* if

(16) $$x_1 \neq x_2 \;\Rightarrow\; F(x_1) \cap F(x_2) = \varnothing$$

(which is contrapositively equivalent to

(17) $$F(x_1) \cap F(x_2) \neq \varnothing \;\Rightarrow\; x_1 = x_2)$$

A semi-single-valued mapping $F : X \multimap Y$ defines a *partition* of its range $\mathrm{ran}(F)$; its distinct values are pairwise disjoint subsets of Y, forming the *blocks* of the partition. It also defines a partition of its domain X: one block is $X \sim \mathrm{cor}(F)$ (which contains all those $x \in X$ for which $F(x) = \varnothing$), and then

$\mathrm{cor}(F)$ is partitioned into blocks that are in one-to-one correspondence with the blocks of $\mathrm{ran}(F)$.

A single-valued mapping $f: X \to Y$ is clearly also a semi-single-valued mapping, and the blocks of the partition of its range $f(X)$ are the singleton sets $\{f(x)\}$. The mapping f induces an equivalence relation R_f on X ($x_1 R_f x_2$ iff $f(x_1) = f(x_2)$), whence defines the single-valued natural mapping, the projection $\pi_f : X \to X/R_f$, which sends $x \in X$ to its R_f-equivalence class $[x]_{R_f} \in X/R_f$ (cf. ML: 2.19–2.21). The single-valued natural mapping $\pi_f : x \mapsto [x]_{R_f}$ may alternatively be formulated as the set-valued mapping $\Pi_f : X \multimap X$ defined by $\Pi_f : x \mapsto [x]_{R_f}$, which sends $x \in X$ to $[x]_{R_f} \subset X$. Π_f is semi-single-valued, since equivalence classes are mutually exclusive.

It is also evident that an injective set-valued mapping is semi-single-valued. Each block of the partition of the *corange* of an injective set-valued mapping is a singleton set. An injective single-valued mapping is an injective set-valued mapping.

2.8 Embedding   For $A \subset X$, the injective set-valued mapping $i : A \multimap X$ defined by $i(x) = \{x\}$ for all $x \in A$ is called the *inclusion map* (or the *embedding*) of A in X. The inclusion map of X in X is called the *identity map* on X, denoted 1_X (whence $1_X : x \mapsto \{x\}$). These match their definitions as (single-valued) mappings (cf. Section 1.13).

As relations, the inclusion map $i : A \multimap X$ is the set $i = \{(x,x) : x \in A\}$ $\subset A \times X (\subset X \times X)$, and the identity map $1_X : X \multimap X$ is the set $1_X = \{(x,x) : x \in X\} \subset X \times X$. Thus each is a member of $\mathrm{P}(X \times X)$ that consists of all the 'diagonal elements' corresponding to the embedded set.

From Sets to Sets

2.9 Definition   Let F be a set-valued mapping from X to Y. If $E \subset X$, the *image* of E under F is defined as the set

(18) $$F(E) = \bigcup_{x \in E} F(x) \subset Y.$$

This is the natural extension of Definition 1.8 of image of a (single-valued) mapping, whence the mapping in (18) is in fact the 'power-set map' $\mathrm{P}F : \mathrm{P}X \to \mathrm{P}Y$. It is also evident that

(19) $$F(X) = \text{ran}(F),$$

whence, in particular, that surjective means $Y = F(X)$.

While the definition of a 'set-to-set' mapping $Pf : PX \to PY$ from a 'point-to-point' mapping $f : X \to Y$ only goes one way (as explained in Section 1.18), the definition of a 'set-to-set' mapping $PF : PX \to PY$ from a 'point-to-set' mapping $F : X \dashv C\, Y$ is reversible. Given a mapping $g : PX \to PY$ of power sets, the assignment, for $x \in X$, $F(x) = g(\{x\})$ naturally defines a set-valued mapping $F : X \dashv C\, Y$ for which $g = Pf$.

The 'set-to-set' mapping $PF : PX \to PY$ has the following properties (*cf.* Theorem 1.19):

2.10 Theorem   *Let $F : X \dashv C\, Y$ be a set-valued mapping and $A, B \subset X$. Then:*

i. $A = \varnothing \;\Rightarrow\; F(A) = \varnothing$.

ii. $A \subset B \;\Rightarrow\; F(A) \subset F(B)$.

iii. $F(A \cup B) = F(A) \cup F(B)$.

iv. $F(A \cap B) \subset F(A) \cap F(B)$.

v. $F(B \sim A) \supset F(B) \sim F(A)$.

Since a (single-valued) mapping is a specialized set-valued mapping through the correspondence (4), whatever properties that set-valued mappings have cannot be contradictory to (but may be weaker than) their analogs for mappings. Properties 2.10.ii–v are identical to their counterparts in Theorem 1.19. But since $F(x) = \varnothing$ is allowed, the implication in property 2.10.i now only goes one way.

2.11 Theorem   *Let $F : X \dashv C\, Y$ be a set-valued mapping. The following are equivalent:*

i.   *F is injective.*

ii. *For all $A, B \subset X$, $A \cap B = \varnothing \;\Rightarrow\; F(A) \cap F(B) = \varnothing$.*

iii. *For all $A, B \subset X$, $F(A \cap B) = F(A) \cap F(B)$.*

iv. *For all $A, B \subset X$, $F(B \sim A) = F(B) \sim F(A)$.*

v.   *For all $A \subset X$, $F(X \sim A) \subset Y \sim F(A)$.*

2.12 Theorem   *Let* $F : X \multimap Y$ *be a set-valued mapping. The following are equivalent:*

i.   *F is surjective.*
ii. *For all* $A \subset X$, $Y \sim F(A) \subset F(X \sim A)$.

Inverse Mapping

2.13 Definition   Given a set-valued mapping $F : X \multimap Y$, its *inverse* is the set-valued mapping $F^{-1} : Y \multimap X$ (equivalently, the relation $F^{-1} \subset Y \times X$) defined by interchanging the ordered components in the graph (5) of F:

$$(20) \qquad F^{-1} = \{(y,x) \in Y \times X : y \in F(x)\} = \{(y,x) \in Y \times X : (x,y) \in F\} \subset Y \times X.$$

A (single-valued) mapping is not necessarily injective, and so its inverse is not necessarily single-valued and hence not (well defined as) a mapping. But the inverse of a set-valued mapping is always a set-valued mapping. Note, however, that F^{-1} is itself a point-to-set mapping (not a 'set-to-point mapping', as a direct reversal-of-roles 'inverse' of a point-to-set mapping would have been), with its value at the point $y \in Y$ defined as the set

$$(21) \qquad\qquad F^{-1}(y) = \{x \in X : (x,y) \in F\} \subset X.$$

Indeed, since both $F(x)$ and $F^{-1}(y)$ are defined by the membership $(x,y) \in F$ (*cf.* (2) and (21)), one trivially has

2.14 Lemma   *Let* $F : X \multimap Y$, $x \in X$, *and* $y \in Y$. *Then*

$$(22) \qquad\qquad y \in F(x) \ \textit{iff} \ x \in F^{-1}(y).$$

While F maps points in X to subsets of Y, the inverse F^{-1} maps points in Y to subsets of X; so the involvements of the sets X and Y in F and F^{-1} are asymmetric. The situation is more evident if one considers the maps in terms of Definition 2.1B:

$$(23) \qquad\qquad F : X \to PY, \qquad F^{-1} : Y \to PX.$$

There is, however, symmetry in corange and range:

$$(24) \qquad \text{cor}(F) = \text{ran}(F^{-1}) = F^{-1}(Y), \qquad F(X) = \text{ran}(F) = \text{cor}(F^{-1}).$$

Note also that

(25) $$\operatorname{dom}(F) = \operatorname{cod}\left(F^{-1}\right) = X, \qquad Y = \operatorname{cod}(F) = \operatorname{dom}\left(F^{-1}\right).$$

And that

(26) $$\left(F^{-1}\right)^{-1} = F.$$

For $F : X \multimap Y$, all the $x \in X$ for which $F(x) = \varnothing$ are not members of $\operatorname{cor}(F)$ and, therefore, not members of $\operatorname{ran}\left(F^{-1}\right)$. In other words, when $X \sim \operatorname{cor}(F) \neq \varnothing$, F^{-1} is not surjective. If $y \in Y \sim \operatorname{ran}(F)$, then $F^{-1}(y) = \varnothing$. Consider the simple example of $F : \{1,2\} \multimap \{p,q\}$ with $F(1) = \{p,q\}$ and $F(2) = \{q\}$; then $F^{-1}(p) = \{1\}$ and $F^{-1}(q) = \{1,2\}$. This F^{-1} is not semi-single-valued and (hence) not injective. Thus, in contrast to an inverse mapping f^{-1} (which is only defined from $\operatorname{ran}(f)$ to X but is both injective and surjective thence, cf. Section 1.15), an inverse set-valued mapping F^{-1} is defined from Y to X, but is not necessarily either injective or surjective.

2.15 Theorem *Let* $F : X \multimap Y$ *and* $F^{-1} : Y \multimap X$ *be its inverse. Then:*

 i. *If F is single-valued, F^{-1} is injective.*
 ii. *If F is injective, F^{-1} is single-valued.*
 iii. *If F is semi-single-valued, F^{-1} is semi-single-valued.*

Inverse Images

If $f : X \to Y$ is a mapping and $E \subset Y$, the inverse image of E under f, the set $f^{-1}(E) = \{x \in X : f(x) \in E\}$, may be considered in two equivalent ways:

 i. As the set $\{x \in X : \{f(x)\} \cap E \neq \varnothing\}$

 ii. As the set $\{x \in X : \{f(x)\} \subset E\}$

When these two sets are interpreted in set-valued mapping terms (recalling that f defines the special singleton-set-valued mapping $x \mapsto \{f(x)\}$), they give two different notions of the inverse image of a set $E \subset Y$:

2.16 Definition   For a set-valued mapping $F : X \multimap Y$ and $E \subset Y$,

i. The *inverse image* of E by F is the set

(27)
$$F^{-1}(E) = \begin{cases} \{x \in X : F(x) \cap E \neq \varnothing\} & \text{if } E \neq \varnothing \\ \varnothing & \text{if } E = \varnothing \end{cases}.$$

ii. The *core* of E by F is the set

(28)
$$F^{+1}(E) = \{x \in X : F(x) \subset E\}.$$

The two notions i and ii coincide (and are identical to the inverse image in Definition 1.12) when the mapping is single-valued, since $F(x) \cap E \neq \varnothing$ iff $F(x) \subset E$ when $F(x)$ is a singleton set.

Note that when $F^{-1} : Y \multimap X$ is considered a set-valued mapping itself (as opposed to its role as the inverse of another set-valued mapping), for $E \subset Y$ the set $F^{-1}(E)$, the image of E under F^{-1}, has already been defined in 2.9. It is the set

(29)
$$F^{-1}(E) = \bigcup_{y \in E} F^{-1}(y) \subset X.$$

One may verify that this defines the same set as in (27), so the notation is consistent. In particular, for $y \in Y$, $F(x) \cap \{y\} \neq \varnothing$ iff $y \in F(x)$ iff $(x, y) \in F$, thus $F^{-1}(\{y\})$ as defined by (27) when $E = \{y\}$ is identical to $F^{-1}(y)$ as defined in (21).

The similarity of the word 'core' to the symbol ' cor ' for corange may lead to confusion, so it is perhaps opportune to clarify here at the outset. For a set-valued mapping $F : X \multimap Y$ and $E \subset Y$, both the corange of F and the core of E by F are subsets of the domain X of F:

(30)
$$\text{cor}(F) \subset X \quad \text{and} \quad F^{+1}(E) \subset X.$$

But there are no general inclusion relations between $\text{cor}(F)$ and $F^{+1}(E)$. Other than having the first three letters of their names in common, corange and core are very different entities: $\text{cor}(\bullet)$, the corange of $\bullet$, accepts one argument F that is a set-valued mapping, whereas $\bullet^{+1}(\bullet)$, the core of $\bullet$ by $\bullet$, accepts two arguments, the first being a set-valued mapping F and the second being a subset E of the mapping's codomain.

The definition of $F^{+1}(E)$ implies that

(31) $F^{+1}(\emptyset) = \{x \in X : F(x) = \emptyset\} = X \sim \mathrm{cor}(F) = \mathrm{dom}(F) \sim \mathrm{cor}(F)$;

i.e., $F^{+1}(\emptyset)$ is the subset of X that contains all those $x \in X$ at which $F(x) = \emptyset$, and it is not necessarily the empty set. Equivalently, (31) says

(32) $\mathrm{cor}(F) = \{x \in X : F(x) \neq \emptyset\} = X \sim F^{+1}(\emptyset) = \mathrm{dom}(F) \sim F^{+1}(\emptyset)$.

Note that for every $E \subset Y$,

(33) $F^{+1}(\emptyset) \subset F^{+1}(E)$,

and

(34) $F^{-1}(E) \subset X \sim F^{+1}(\emptyset)$.

This last inclusion says that $F^{-1}(E) \cap F^{+1}(\emptyset) = \emptyset$, which means if $x \in F^{-1}(E)$, then $F(x) \neq \emptyset$.

Consider the simple example of $F : \{1,2\} \rightrightarrows \{p,q\}$ with $F(1) = \{p,q\}$ and $F(2) = \emptyset$; then $\mathrm{cor}(F) = \{1\}$, $F^{-1}(\{p\}) = \{1\}$, and $F^{+1}(\{p\}) = \{2\}$. This shows that in general there are no inclusion relations between $\mathrm{cor}(F)$ and $F^{+1}(E)$ and between $F^{-1}(E)$ and $F^{+1}(E)$.

The same authors who define the domain of F as $\{x \in X : F(x) \neq \emptyset\}$ (i.e., my $\mathrm{cor}(F)$) also define their alternate inverse (sometimes called *upper inverse*) accordingly, for $E \subset Y$, as

(35) $F^{\wedge 1}(E) = \begin{cases} \{x \in X : F(x) \neq \emptyset \text{ and } F(x) \subset E\} & \text{if } E \neq \emptyset \\ \emptyset & \text{if } E = \emptyset \end{cases}$.

This puts, for all $E \subset Y$,

(36) $F^{\wedge 1}(E) \subset X \sim F^{+1}(\emptyset) = \mathrm{cor}(F)$.

One sees that

(37) $F^{+1}(E) = F^{\wedge 1}(E) \cup F^{+1}(\emptyset)$ and $F^{\wedge 1}(E) \cap F^{+1}(\emptyset) = \emptyset$

(i.e., $F^{+1}(E)$ is the union of the disjoint sets $F^{\wedge 1}(E)$ and $F^{+1}(\varnothing)$), and

$$(38) \qquad F^{\wedge 1}(E) \subset F^{-1}(E).$$

Also

$$(39) \qquad F^{-1}(E) \cap F^{+1}(\varnothing) = \varnothing .$$

In particular,

$$(40) \qquad F^{-1}(Y) = F^{\wedge 1}(Y) = X \sim F^{+1}(\varnothing) = \mathrm{cor}(F)$$

and

$$(41) \qquad F^{+1}(Y) = X .$$

2.17 Lemma   *For a set-valued mapping $F: X \multimap Y$ and $E \subset Y$,*

$$(42) \qquad F^{-1}(Y \sim E) = X \sim F^{+1}(E);$$

$$(43) \qquad F^{+1}(Y \sim E) = X \sim F^{-1}(E).$$

With the identities (32) and (37), one has

2.18 Corollary   *For a set-valued mapping $F: X \multimap Y$ and $E \subset Y$,*

$$(44) \qquad F^{-1}(Y \sim E) = \mathrm{cor}(F) \sim F^{\wedge 1}(E);$$

$$(45) \qquad F^{\wedge 1}(Y \sim E) = \mathrm{cor}(F) \sim F^{-1}(E).$$

Note that among the three varieties of 'inverse images' that I have defined, inverse image $F^{-1}(E)$, core $F^{+1}(E)$, and upper inverse $F^{\wedge 1}(E)$, only the first is associated with an 'inverse mapping', viz., $F^{-1}: Y \multimap X$, with

$$(46) \qquad F^{-1}(y) = F^{-1}(\{y\}) = \{x \in X : F(x) \cap \{y\} \neq \varnothing\} = \{x \in X : y \in F(x)\} .$$

While one may similarly define $F^{+1}(y)$ and $F^{\wedge 1}(y)$,

$$(47) \qquad F^{+1}(y) = F^{+1}(\{y\}) = \{x \in X : F(x) \subset \{y\}\},$$

(48) $F^{\wedge 1}(y) = F^{\wedge 1}(\{y\}) = \{x \in X : F(x) \neq \varnothing \text{ and } F(x) \subset \{y\}\},$

the resulting mappings are not very useful. The restriction (48) means $F^{\wedge 1}(y)$ would contain only those $x \in X$ for which $F(x)$ is the singleton set $\{y\}$, and (47) just means $F^{+1}(y)$ would contain all those $x \in X$ for which $F(x)$ is either the empty set $\varnothing$ or the singleton set $\{y\}$, i.e., $F^{+1}(y) = F^{+1}(\varnothing) \cup F^{\wedge 1}(y)$.

2.19 Theorem   *Let* $F : X \multimap Y$ *and* $A, B \subset Y$. *Then:*

i. $A = \varnothing \;\Rightarrow\; F^{-1}(A) = \varnothing$.

ii. $A \subset B \;\Rightarrow\; F^{-1}(A) \subset F^{-1}(B)$.

iii. $F^{-1}(A \cup B) = F^{-1}(A) \cup F^{-1}(B)$.

iv. $F^{-1}(A \cap B) \subset F^{-1}(A) \cap F^{-1}(B)$.

v. $F^{-1}(B \sim A) \supset F^{-1}(B) \sim F^{-1}(A)$.

Compare this with Theorem 2.10 for the corresponding properties of F. Note that the two theorems are in fact the same (with the obvious corresponding changes due to the replacement of $F : X \multimap Y$ in 2.10 by $F^{-1} : Y \multimap X$ in 2.19). I put Theorem 2.19 here to emphasize the point that $F^{-1} : Y \multimap X$ is a 'general' set-valued mapping like any other, without any inherent special properties. This fact is different from the case of (single-valued) mappings, for which f^{-1} only exists when f is injective, and this specialization of f gives f^{-1} stronger properties that f^{-1} is necessarily bijective. Compare Theorems 2.10 and 2.19 with their counterparts for f and f^{-1}, Theorems 1.19 and 1.20.

Iterated Mappings of Sets

Theorem 1.21 lists three properties of the combination of a mapping $f : X \to Y$ and its (not-necessarily-a-mapping) inverse f^{-1} when they map sets. I shall examine their counterparts for set-valued mappings.

2.20 First Combination   Theorem 1.21.i says that for $A \subset X$, one has $A \subset f^{-1}(f(A))$. But there is no corresponding property $A \subset F^{-1}(F(A))$ for a set-valued mapping $F : X \multimap Y$. Consider the simple example of $F : \{1, 2\} \multimap$

$\{p,q\}$ with $F(1)=\{p\}$ and $F(2)=\varnothing$; then $F^{-1}(F(\{1,2\}))=F^{-1}(\{p\})$ $=\{1\}$, so $\{1,2\}\not\subset F^{-1}(F(\{1,2\}))$.

Suppose $F:X\multimap Y$ and $A\subset X$. Recall ((31) above) $F^{+1}(\varnothing)$ is the subset of X that contains all those $x\in X$ at which $F(x)=\varnothing$. It follows from (39) above that $F^{-1}(F(A))\cap F^{+1}(\varnothing) = \varnothing$; i.e., $F^{-1}(F(A))\subset \text{cor}(F)$. Now $F^{-1}(F(A))$ is the set $\{x\in X:F(x)\cap F(A)\neq\varnothing\}$, so if $x\in A$ and $F(x)\neq\varnothing$, then $x\in F^{-1}(F(A))$. Thus $A\sim F^{+1}(\varnothing)\subset F^{-1}(F(A))$. Stated otherwise, if one restricts to subsets $A\subset \text{cor}(F)=X\sim F^{+1}(\varnothing)$, then one does have $A\subset F^{-1}(F(A))$. Conversely, since $F^{-1}(F(A))\subset \text{cor}(F)$, if $A\subset F^{-1}(F(A))$, then *a fortiori* $A\subset \text{cor}(F)$. Thus

2.21 Lemma   Let $F:X\multimap Y$. Then $A\subset F^{-1}(F(A))$ *iff* $A\subset \text{cor}(F)$.

For $F:X\multimap Y$, $G(x)=F^{-1}(F(x))$ is a set-valued mapping from X to X. Lemma 2.21 says that if $x\in \text{cor}(F)$ (i.e., if $F(x)\neq\varnothing$), then $x\in F^{-1}(F(x))$, which means the 'diagonal element' $(x,x)\in G$. Let $i:\text{cor}(F)\multimap X$ be the inclusion map of $\text{cor}(F)$ in X (Section 2.8).   Then as a relation $i=\{(x,x):x\in \text{cor}(F)\}\subset \text{cor}(F)\times X(\subset X\times X)$, whence $i\subset G$.

Subsets $A\subset X$ for which $A=F^{-1}(F(A))$ are special:

2.22 Definition   Let $F:X\multimap Y$. A subset $A\subset X$ for which $A=F^{-1}(F(A))$ is called a *stable subset* (of X under F).

It follows from Lemma 2.21 that a stable subset must be a subset of $\text{cor}(F)$.

2.23 Theorem   Let $F:X\multimap Y$. *The stable subsets form a complemented lattice* $\mathfrak{S}$ *(a complemented sublattice of the power-set lattice* $\mathrm{P}(\text{cor}(F))$ *).*

Proof   Note that $F^{-1}(F(\varnothing))=\varnothing$ and $F^{-1}(F(\text{cor}(F)))=F^{-1}(\text{ran}(F))$ $=\text{cor}(F)$, so $\varnothing\in\mathfrak{S}$ and $\text{cor}(F)\in\mathfrak{S}$ (respectively the least element and the greatest element of $\mathfrak{S}$).

Let $A \subset \text{cor}(F)$, whence $A \subset F^{-1}(F(A))$ by Lemma 2.21. A nonempty $F^{-1}(F(A)) \sim A$ means the existence of an element $x \in F^{-1}(F(A)) \sim A$. This element x must be in $\text{cor}(F) \sim A$, which means $F(x) \cap F(\text{cor}(F) \sim A) \neq \varnothing$. At the same time, $x \in F^{-1}(F(A)) \sim A$, so a *fortiori* $x \in F^{-1}(F(A))$, which means $F(x) \cap F(A) \neq \varnothing$. In other words, $A = F^{-1}(F(A))$ iff there is no $x \in \text{cor}(F)$ such that $F(x) \cap F(A) \neq \varnothing$ and $F(x) \cap F(\text{cor}(F) \sim A) = \varnothing$. But this equivalent condition is the same when A is replaced by $\text{cor}(F) \sim A$, since $\text{cor}(F) \sim (\text{cor}(F) \sim A) = A$, whence it also defines the conditions under which $\text{cor}(F) \sim A = F^{-1}(F(\text{cor}(F) \sim A))$. Thus $A = F^{-1}(F(A))$ iff $\text{cor}(F) \sim A = F^{-1}(F(\text{cor}(F) \sim A))$, and this says $A \in \mathfrak{S}$ iff $\text{cor}(F) \sim A \in \mathfrak{S}$. $\mathfrak{S}$ is therefore complemented.

Let $A, B \in \mathfrak{S}$. Then, using Theorem 2.10.iii and Theorem 2.19.iii,

$$(49) \qquad F^{-1}(F(A \cup B)) = F^{-1}(F(A)) \cup F^{-1}(F(B)) = A \cup B,$$

so $A \cup B \in \mathfrak{S}$. Since $\text{cor}(F) \sim (A \cap B) = (\text{cor}(F) \sim A) \cup (\text{cor}(F) \sim B) \in \mathfrak{S}$, one also has $A \cap B \in \mathfrak{S}$.   □

Theorem 1.22 says that a mapping $f : X \to Y$ is injective iff $A = f^{-1}(f(A))$ for all $A \subset X$. Correspondingly, one has

2.24 Lemma   *A set-valued mapping $F : X \multimap Y$ is injective iff $A = F^{-1}(F(A))$ for all $A \subset \text{cor}(F)$.*

When a set-valued mapping $F : X \multimap Y$ is injective, every subset of $\text{cor}(F)$ is stable; the complemented lattice $\mathfrak{S}$ of stable subsets is thus all of $P(\text{cor}(F))$.

An injective mapping $f : X \to Y$ means $f^{-1} \circ f = 1_X$, the identity mapping $x \mapsto x$ on the domain X. But an injective set-valued mapping $F : X \multimap Y$ means

$$(50) \qquad F^{-1}(F(x)) = \begin{cases} \{x\} & \text{if } x \in \text{cor}(F) \\ \varnothing & \text{if } F(x) = \varnothing \end{cases},$$

i.e., $x \mapsto F^{-1}(F(x))$ is a disjoint union, the concatenation of the inclusion map i of $\mathrm{cor}(F)$ in X and the constant empty-set-valued mapping $\varnothing$ on $F^{+1}(\varnothing) = X \sim \mathrm{cor}(F)$. So even for an injective F, the combination $G(x) = F^{-1}(F(x))$ is still not quite the identity mapping on X (unless $F^{+1}(\varnothing) = \varnothing$). When i and G are considered as subsets of $X \times X$, $G = i = \{(x,x) : x \in A\}$ (but not necessarily $G = 1_X = \{(x,x) : x \in X\}$).

2.25 Second Combination   Given a mapping $f : X \to Y$ and $B \subset Y$, one has $B \supset f(f^{-1}(B))$ (Theorem 1.21.ii). But there is no containment relation between B and $F(F^{-1}(B))$ for a set-valued mapping $F : X \multimap Y$. Consider the example $F : \{1,2\} \multimap \{p,q,r\}$ with $F(1) = \{p,q\}$ and $F(2) = \varnothing$; then $F(F^{-1}(\{q,r\}))$ $= F(\{1\}) = \{p,q\}$. This time, even a restriction to $B \subset \mathrm{ran}(F)$ (dual to $A \subset$ $\mathrm{cor}(F)$ in Lemma 2.21) does not help: in the example, $\{p\} \not\supset F(F^{-1}(\{p\})) =$ $F(\{1\}) = \{p,q\}$. Neither does the specialization to surjections: Theorem 1.23 says that a mapping $f : X \to Y$ is surjective iff $B = f(f^{-1}(B))$ for all $B \subset Y$. But the same F in my example is a surjective set-valued mapping from $\{1,2\}$ onto $\{p,q\}$, and still $\{p\} \neq F(F^{-1}(\{p\})) = F(\{1\}) = \{p,q\}$.

Since neither $x \mapsto F^{-1}(F(x))$ nor $y \mapsto F(F^{-1}(y))$ is necessarily the identity mapping on its respective domain, one must understand the usage of the term '*inverse* set-valued mapping' with this in mind: it is not the usual algebraic definition in connection with a 'reversal entity for the recovery of the identity'. For this reason, some authors call $F^{-1} : Y \multimap X$ the 'converse' of $F : X \multimap Y$ instead of the 'inverse'.

2.26 Third Combination   Given a mapping $f : X \to Y$, $A \subset X$, and $B \subset Y$, one has $f(A \cap f^{-1}(B)) = f(A) \cap B$ (Theorem 1.21.iii).

Consider the example $F : \{1,2\} \multimap \{p,q\}$ with $F(1) = \{p,q\}$ and $F(2)$ $= \{q\}$; then $F(\{1\} \cap F^{-1}(\{p\})) = F(\{1\} \cap \{1\}) = \{p,q\}$, but $F(\{1\}) \cap \{p\}$ $= \{p,q\} \cap \{p\} = \{p\}$. So they are not equal. But one does have inclusion:

2.27 Theorem   Let $F : X \multimap Y$, $A \subset X$, and $B \subset Y$. Then $F(A \cap F^{-1}(B))$ $\supset F(A) \cap B$.

Operations on Set-Valued Mappings

2.28 Definition   If $F : X \multimap Y$ and $G : X \multimap Y$ are two set-valued mappings, then:

i. Their *union* is the mapping $F \cup G : X \multimap Y$ defined by
$$(F \cup G)(x) = F(x) \cup G(x).$$

ii. Their *intersection* is the mapping $F \cap G : X \multimap Y$ defined by
$$(F \cap G)(x) = F(x) \cap G(x).$$

iii. Their *Cartesian product* is the mapping $F \times G : X \multimap Y \times Y$ defined by
$$(F \times G)(x) = F(x) \times G(x).$$

2.29 Theorem   Let $F : X \multimap Y$ and $G : X \multimap Y$. Then, for $A \subset X$:

i. $(F \cup G)(A) = F(A) \cup G(A).$

ii. $(F \cap G)(A) \subset F(A) \cap G(A).$

iii. $(F \times G)(A) \subset F(A) \times G(A).$

Recall (Section 2.4) that $F : X \multimap Y$ is a constant (set-valued) mapping if there exists a subset C of Y such that $F(x) = C$ for all $x \in X$. This implies $F(A) = C$ for all $A \subset X$.

2.30 Corollary   Let $F : X \multimap Y$ be a constant mapping. Let $G : X \multimap Y$ and $A \subset X$. Then $(F \cap G)(A) = F(A) \cap G(A).$

2.31 Theorem   If both $F : X \multimap Y$ and $G : X \multimap Y$ are semi-single-valued, then the set-valued mappings $F \cap G : X \multimap Y$ and $F \times G : X \multimap Y \times Y$ are semi-single-valued.

2.32 Theorem   If one of $F : X \multimap Y$ and $G : X \multimap Y$ is injective, then the set-valued mappings $F \cap G : X \multimap Y$ and $F \times G : X \multimap Y \times Y$ are injective.

3
Principles of Set-Valued Mappings

> This is the law pertaining to land animal and bird and every living creature that moves through the waters and every creature that swarms upon the earth.
>
> — *Leviticus* 11:46

Explanatio: Relational Diagram

3.1 Four Causes in a Mapping   When a mapping $f : X \to Y$ is represented in the element-chasing version $f : a \mapsto b$ (where $a \in X$ and $b = f(a) \in Y$), its relational diagram may be drawn as a network with three *nodes* and two *directed edges*, that is, a directed graph (or *digraph* for short):

(1)
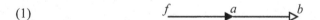

The reader is referred to Chapter 5 of *ML* for the details of the symbology, which I shall not repeat here. A terse summary is that in this arrow-diagram representation (1), Aristotle's four causes appear as components. The efficient cause, the processor of entailment, is explicitly shown as the solid-headed arrow (the former being at the tail of the latter). The final cause, the target of entailment, is explicitly shown as the hollow-headed arrow (the former being at the head of the latter). The other two causes, material and final, are included implicitly: the material cause, the input, is at the tail of the hollow-headed arrow (and also at the head of the solid-headed arrow); the formal cause, the *morphé* of the morphism, is the ordered pair of arrows, the mathematical structure of the two-arrow digraph (1) itself. The embedding of the four causes as components of diagram (1) is succinctly summarized in

A.H. Louie, *The Reflection of Life*, IFSR International Series on Systems Science and Engineering 29, DOI 10.1007/978-1-4614-6928-5_3, © Springer Science+Business Media New York 2013

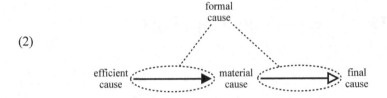

(2)

(which is diagram (10) in *ML*: Chapter 5).

3.2 Arrow Diagram of a Set-Valued Mapping   For a set-valued mapping $F : X \multimap Y$ in its element-chasing version $F : a \mapsto B$ (where $a \in X$ and $B = F(a) \subset Y$), its relational diagram may be drawn thus

(3) $$ F \xrightarrow{\quad\quad} a \xrightarrow{\quad\quad} B $$

The same representation as an ordered pair of solid-headed arrow and hollow-headed arrow suffices. Context determines the nature of the final cause: whether it is an 'element', a 'set', or some other entity.

3.3 Material Connections   Note that for the three operations $F \cup G$, $F \cap G$, and $F \times G$ in Definition 2.28, the two original set-valued mappings involved, $F : X \multimap Y$ and $G : X \multimap Y$, must have the same domain, because the 'combined' mappings are defined on this same domain. The relational diagrams of F and G therefore meet at the common node a that is the material cause of both set-valued mappings:

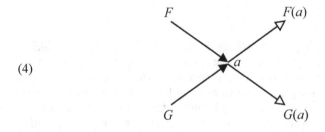

(4)

The 'combined' set-valued mappings are thus

(5) $$ F*G \xrightarrow{\quad a \quad} F(a)*G(a) $$

(where $*$ is one of $\cup$, $\cap$, or $\times$). Note that there is no 'relaying' of any component involved in these 'combined' set-valued mappings; F and G act on $a \in X$ simultaneously and the 'combined' set-valued mapping yields $F(a) * G(a)$.

While I have used the same codomains for F and G in Definition 2.28, that is not a requisite; they may be different (with the obvious modifications in the definitions of the combined maps). But if they are 'too different', the maps $F \cup G$ and $F \cap G$ would not be too interesting. For example, if the codomains are disjoint, then $F \cup G$ is just a disjoint union, and $F \cap G$ is the constant mapping $\varnothing : X \multimap Y$ that sends everything to the empty set (*cf.* Section 2.18).

Sequential Composition

When the final cause (output) of a mapping is relayed to another mapping as the latter's material cause (input), the two mappings compose *sequentially*. Just as there are two different notions of the inverse image, there are likewise two different ways to define 'sequential composition' for set-valued mappings. (It will turn out that these two notions of sequential composition are related correspondingly to the two notions of the inverse image, as we shall see presently.)

3.4 Definition   Let $F : X \multimap Y$ and $G : Z \multimap X$ be set-valued mappings.

i. The *sequential composition* (or simply *composite* or *product*) is the set-valued mapping $F \circ G : Z \multimap Y$ defined by, for $z \in Z$,

(6)
$$(F \circ G)(z) = \bigcup_{x \in G(z)} F(x) \subset Y.$$

ii. The *square product* is the set-valued mapping $F \square G : Z \multimap Y$ defined by, for $z \in Z$,

(7)
$$(F \square G)(z) = \bigcap_{x \in G(z)} F(x) \subset Y.$$

Note that *the codomain of G is the domain of F*, enabling the two compositions that are the sequence "G followed by F". Also note the symbols used for the two binary operations: for sequential composition in (6), it is the standard 'small circle' $\circ$ of 'composite'; for square product in (7), it is a 'small square' $\square$.

One readily verifies that

(8)
$$(F \circ G)(z) = F(G(z)).$$

Note that in the iteration on the right-hand side of (8), G takes the point $z \in Z$ to the set $G(z) \subset X$, then the 'power set map' F (as in Definition 2.9) relays the set $G(z) \subset X$ to the set $F(G(z)) \subset Y$. On the left-hand side, the sequential composition $F \circ G$ is a set-valued mapping that combines these two steps into one, taking the point $z \in Z$ to the set $(F \circ G)(z)$ (as defined in (6)). So equality (8) is not a tautology, but a statement that the two sets on either side, while defined differently, are in fact identical. Stated otherwise, the efficient causes on the two sides of (8) take separate and different paths, but beginning with the same material cause z, they reach the same final cause at the end.

Concerning the difference between the two products, one may consider that the sequential composition $(F \circ G)(z)$ traces *at least one* path of the element $z \in Z$ as it is mapped by G into X and then by F into Y, while the square product $(F \square G)(z)$ traces *all* such paths.

It is trivial to verify that both the sequential composition and the square product are associative.

Since $A \cap B \subset A \cup B$, one may be tempted to conclude from the definitions of the two products that $(F \square G)(z) \subset (F \circ G)(z)$. But this is not the case, because the union and intersection may be taken over an empty set, viz. when $G(z) = \varnothing$. In a lattice L, the least element and greatest element (when they exist) are $\inf L = \sup \varnothing$ and $\sup L = \inf \varnothing$. For the lattice PY, therefore, one has $\varnothing = \inf PY = \bigcup_{\varnothing}$ and $Y = \sup PY = \bigcap_{\varnothing}$ (*cf.* ML: 1.28). Thus when $G(z) = \varnothing$,

$$(F \circ G)(z) = \bigcup_{x \in \varnothing} F(x) = \varnothing \quad \text{and} \quad (F \square G)(z) = \bigcap_{x \in \varnothing} F(x) = Y, \quad \text{whence} \quad (F \square G)(z)$$

$$\not\subset (F \circ G)(z).$$

When G is single valued, $x \in G(z)$ iff $G(z) = \{x\}$, whence the union in (6) and intersection in (7) are both taken over the same single element, hence identical. For standard (single-valued) mappings, the two products i and ii therefore coincide (and are identical to the standard sequential composition of mappings).

The relationships of the two products with inverse image and core (Definition 2.16) are as follows:

3.5 Lemma   *Let* $F: X \multimap Y$ *and* $G: Z \multimap X$ *be set-valued mappings. If* $E \subset Y$, *then*

(9) $$(F \circ G)^{-1}(E) = G^{-1}(F^{-1}(E));$$

(10) $$(F \square G)^{+1}(E) = G^{+1}(F^{+1}(E)).$$

The connection between inversion and sequential composition of set-valued mappings is exactly isomorphic to that of (single-valued) mappings:

3.6 Theorem   *Let* $F : X \multimap Y$ *and* $G : Z \multimap X$ *be set-valued mappings. If* $y \in Y$, *then*

$$(11) \qquad (F \circ G)^{-1}(y) = G^{-1}\left(F^{-1}(y)\right) = \left(G^{-1} \circ F^{-1}\right)(y).$$

Taking Lemma 2.14 into account, one has the following sequence of equivalences:

3.7 Corollary   *Let* $F : X \multimap Y$ *and* $G : Z \multimap X$ *be set-valued mappings. Then*

$$y \in (F \circ G)(z) \Leftrightarrow z \in (F \circ G)^{-1}(y) \Leftrightarrow z \in G^{-1}\left(F^{-1}(y)\right)$$
$$\Leftrightarrow G(z) \cap F^{-1}(y) \neq \varnothing$$

$(12) \qquad\quad \Leftrightarrow \exists x \in X : x \in G(z) \ and \ x \in F^{-1}(y)$

$$\Leftrightarrow \exists x \in X : x \in G(z) \ and \ y \in F(x)$$
$$\Leftrightarrow \exists x \in X : (z, x) \in G \ and \ (x, y) \in F.$$

The final statement in the sequence is precisely the one used in the following:

3.8 Definition   Let $F \subset X \times Y$ and $G \subset Z \times X$ be relations. Their *relative product* $F \circ G$ is the set of all ordered pairs $(z, y) \in Z \times Y$ for which there exists an $x \in X$ with $(z, x) \in G$ and $(x, y) \in F$:

$$(13) \qquad (F \circ G) = \left\{ (z, y) \in Z \times Y : \exists x \in X : (z, x) \in G \text{ and } (x, y) \in F \right\}$$

(*cf. ML*: 2.35).   Thus the sequential composition (i.e. product) of set-valued mappings is identical to their relative product as relations.

The connection between inversion and square product of set-valued mappings takes the following form:

3.9 Theorem   *Let* $F : X \multimap Y$ *and* $G : Z \multimap X$ *be set-valued mappings. If* $y \in Y$, *then*

$$(14) \qquad\qquad (F \square G)^{-1}(y) = G^{+1}\left(F^{-1}(y)\right).$$

Note the G^{+1} and F^{-1} appearing on the right-hand side of (14); they are not typographical errors. Also note that while for sequential composition one has the nice equality $\left(F\circ G\right)^{-1}(y) = \left(G^{-1}\circ F^{-1}\right)(y)$, for square product there is no inclusion relation between $\left(F\circ G\right)^{-1}(y)$ and $\left(G^{-1}\circ F^{-1}\right)(y)$. Instead, one has the slightly more complicated

3.10 Theorem   *Let* $F:X \multimap Y$ *and* $G:Z \multimap X$ *be set-valued mappings and* $y\in Y$. *Then*

(15) $$z\in\left(F\circ G\right)^{-1}(y) \quad \textit{iff} \quad G(z)\subset F^{-1}(y);$$

(16) $$z\in\left(G^{-1}\circ F^{-1}\right)(y) \quad \textit{iff} \quad F^{-1}(y)\subset G(z).$$

Thus

(17) $$G(z) = F^{-1}(y) \quad \textit{iff} \quad z\in\left(G^{-1}\circ F^{-1}\right)(y) \cap \left(F\circ G\right)^{-1}(y).$$

Contrast (17) for square product with

(18) $$G(z)\cap F^{-1}(y) \neq \varnothing \quad \textit{iff} \quad z\in\left(F\circ G\right)^{-1}(y)$$

for sequential composition.

One sees from the results of the several previous sections that the sequential composition $F\circ G$ is much better behaved algebraically than the square product $F\square G$. The former is used in most circumstances. But occasionally it is useful to trace *all* paths of an element through composite mappings instead of just *some* paths, then the square product will be the better equipment. (One may anticipate and consider how this is so when these mappings are connected in (M,R)-networks.)

Category of Relations

3.11 The Category Rel   The category in which the collection of objects is the collection of all sets (in a suitably naive universe of small sets) and where morphisms are (binary) relations (as in Definition 1.3) is denoted **Rel**. Given two sets X and Y, the hom-set $\mathbf{Rel}(X,Y)$ of *all* relations between X and Y is thus the power set $\mathsf{P}(X\times Y)$. Equivalently (*cf.* Definition 2.1A), a **Rel**-morphism $F\in\mathbf{Rel}(X,Y)$ is a set-valued mapping $F:X \multimap Y$ from X to Y. The domain $\mathrm{dom}(F)=X$ and the codomain $\mathrm{cod}(F)=Y$, predicated to be determined uniquely by F under a category axiom, are as defined in 2.3 and 2.6. (This is the category-theoretic reason for my choosing $\mathrm{dom}(F) = X$ while coining the new term 'corange' for $\mathrm{cor}(F) = \{x \in X : F(x) \neq \varnothing\}$.)

For relations $F \subset X \times Y$ and $G \subset Z \times X$, their composite is their *relative product* $F \circ G \subset Z \times Y$ (Definition 3.8). Equivalently (Corollary 3.7), for set-valued mappings $F : X \multimap Y$ and $G : Z \multimap X$, their composite is their sequential composition $F \circ G : Z \multimap Y$ (Definition 3.4.i).

The identity $1_X \in \mathbf{Rel}(X,X)$ is the set-valued mapping $1_X : x \mapsto \{x\}$ (Section 2.8).

The category **Set** of sets and (single-valued) mappings is thus a *subcategory* of **Rel**: the objects of the two categories are identical, a **Set**-morphism is a **Rel**-morphism, and compositions as **Rel**-morphisms and as **Set**-morphisms are the same for mappings. A mapping is a special relation, however, so $\mathbf{Set}(X,Y) \subset \mathbf{Rel}(X,Y)$ is a proper inclusion, whence **Set** is not a full subcategory of **Rel**.

I often employ non-full subcategories of **Rel** itself, and I use $H(X,Y)$ for appropriate subsets of $P(X \times Y) = \mathbf{Rel}(X,Y)$ under consideration (e.g. when mappings $F : X \multimap Y$ represent metabolic functions). These subsets $H(X,Y)$ themselves, of course, still have to satisfy the category axioms. This is the same notation used in **Set** when $H(X,Y) \subset Y^X = \mathbf{Set}(X,Y)$, but the context will determine the category hence the nature of the mappings at hand.

3.12 Inverse Functor   The *inverse functor* is the contravariant functor $Inv : \mathbf{Rel} \to \mathbf{Rel}$ (or equivalently the covariant functor $\overline{Inv} : \mathbf{Rel} \to \mathbf{Rel}^{op}$) that sends each set X to itself and each set-valued mapping $F : X \multimap Y$ to its inverse $F^{-1} : Y \multimap X$. The contravariance is due to

$$(19) \qquad (F \circ G)^{-1}(y) = (G^{-1} \circ F^{-1})(y) \quad \text{for all } y \in Y$$

(*cf.* (9) above), whence

$$(20) \qquad Inv(F \circ G) = Inv(G) \circ Inv(F).$$

Since

$$(21) \qquad 1_X^{-1} = 1_X ,$$

trivially

$$(22) \qquad Inv(1_X) = 1_{Inv(X)} .$$

Also

(23) $$\left(F^{-1}\right)^{-1} = F$$

implies that

(24) $$Inv \circ Inv = I_{\mathbf{Rel}}$$

(where $I_{\mathbf{Rel}}$ is the identity functor on **Rel**; *cf. ML*: A.12(i)) and *Inv* is, naturally, an *involution*.

Note that there is no similar inverse functor for **Set**, since the 'inverse' f^{-1} of a (single-valued) mapping f is not necessarily itself single-valued (*cf.* Section 1.15). Even when a mapping $f : X \to Y$ is injective whence its inverse f^{-1} is a mapping, the necessarily bijective inverse mapping $f^{-1} : f(X) \to X$ is only defined on the range $\operatorname{ran}(f) = f(X) \subset Y$: for $y \in Y \sim f(X)$, $f^{-1}(y)$ is undefined. (One only has $f^{-1} : Y \to X$ iff f is surjective as well.) Contrast this with $F^{-1} : Y \multimap X$, for which if $y \in Y \sim F(X)$, $F^{-1}(y)$ is defined and takes the value $F^{-1}(y) = \varnothing$.

Relations have more elaborate properties when the domain and the codomain coincide. I now concentrate on the hom-set $\mathbf{Rel}(X, X)$. Let us revisit the important crew members from *ML*—equivalence relations, partial orders, and graphs—and see how they can all be naturally formulated within the theory of set-valued mappings.

Equivalence Relations

3.13 Definition   If X is a set and $R \subset X \times X$, one says that R is a *relation on* X and writes $x R y$ instead of $(x, y) \in R$.

A relation on X is sometimes, for emphasis, called a *binary relation* on X , since the concept of $R \subset X \times X = X^2$ may be extended to that of an *n*-ary relation $R \subset X^n$ on X . A (binary) relation R on X is thus a set-valued mapping $R : X \multimap X$, that is, $R \in \mathbf{Rel}(X, X)$, with $y \in R(x)$ iff $x R y$.

3.14 Definition   A relation R on a set X is said to be:

(*r*) *Reflexive* if for all $x \in X$, $x R x$
(*s*) *Symmetric* if for all $x, y \in X$, $x R y$ implies $y R x$

(a) *Antisymmetric* if for all $x, y \in X$, $x R y$ and $y R x$ imply $x = y$

(t) *Transitive* if for all $x, y, z \in X$, $x R y$ and $y R z$ imply $x R z$

3.15 Definition The *equality* (or *identity*) relation I on X is defined by $x I y$ if $x = y$.

As a subset of $X \times X$, I is the *diagonal* $I = \{(x, x) : x \in X\}$. As a set-valued mapping $I : X \to X$, it is identical to the identity 1_X in the hom-set $\mathbf{Rel}(X, X)$ (*cf.* Section 3.11), that is, for all $x \in X$, $I(x) = 1_X(x) = \{x\}$. I shall use the two notations I and 1_X interchangeably.

The properties in Definition 3.14 may be formulated in terms of a set-valued mapping $R : X \to X$ as follows:

3.16 Lemma *For a set-valued mapping $R : X \to X$:*

(r) *Reflexive means for all $x \in X$, $x \in R(x)$; thus*
 R *is reflexive iff $I \subset R$.*

(s) *Symmetric means for all $x, y \in X$, $y \in R(x)$ implies $x \in R(y)$ whence*
 $y \in R^{-1}(x)$; *thus*
 R *is symmetric iff $R \subset R^{-1}$*
 (which implies $(R)^{-1} \subset (R^{-1})^{-1}$ whence also that $R^{-1} \subset R$; thus, a fortiori,
 R *is symmetric iff $R = R^{-1}$).*

(a) *Antisymmetric means for all $x, y \in X$, $y \in R(x)$ and $x \in R(y)$ imply*
 $x = y$; *thus*
 R *is antisymmetric iff $R \cap R^{-1} \subset I$.*

(t) *Transitive means for all $x, y, z \in X$, $y \in R(x)$ and $z \in R(y)$ imply*
 $z \in R(x)$; *thus*
 R *is transitive iff $R \circ R \subset R$.*

Note that the defining statements of symmetry, antisymmetry, and transitivity are all conditionals. It is possible for the antecedents to be vacuously true, that is, for no elements to satisfy the requirements of the if clauses. Thus, the empty-set-valued mapping $\varnothing : X \to X$ (which is also the constant empty-set-valued mapping; *cf.* Sections 1.3 and 2.5) is symmetric, antisymmetric, and transitive; one sees that $\varnothing$ satisfies the takes of the lemma (noting that $\varnothing^{-1} = \varnothing$): (s) $\varnothing \subset \varnothing^{-1}$, (a) $\varnothing \cap \varnothing^{-1} \subset I$, and (t) $\varnothing \circ \varnothing \subset \varnothing$.

The defining statement of reflexivity, however, is a declaration of existence. Since reflexivity (r) means for each $x \in X$, $x \in R(x)$, a reflexive R implies $R(x) \neq \varnothing$ and $R^{-1}(x) \neq \varnothing$, whence $\text{cor}(R) = \text{ran}(R) = X$. The empty-set-valued mapping $\varnothing$ is, thus in particular, not reflexive: $\neg(r)$ $I \not\subset \varnothing$ (except when $X = \varnothing$).

The universal relation $U = X \times X : X \dashv\subset X$ is reflexive, symmetric, and transitive, but it is evidently not antisymmetric (unless $X = \varnothing$ or X is a singleton set).

3.17 Equivalence Relation   Let R be an *equivalence relation* on the set X, that is, the set-valued mapping $R : X \dashv\subset X$ is (r) reflexive, (s) symmetric, and (t) transitive (*cf. ML*: 1.11). For each $x \in X$, $R(x)$ is the equivalence class of x determined by R:

$$(25) \qquad R(x) = [x]_R = \{ y \in X : (x, y) \in R \}.$$

Again, reflexivity (r) implies that for each $x \in X$, $x \in [x]_R \neq \varnothing$.

For an equivalence relation R, $R \circ R = R$, whence *idempotence*.

3.18 Definition   Given a set-valued mapping $F : X \dashv\subset W$, one calls two elements $x, y \in X$ *F-related* when $F(x) = F(y)$ and denotes this relation by R_F; that is,

$$(26) \qquad x \; R_F \; y \quad \text{iff} \quad F(x) = F(y).$$

Then R_F is an equivalence relation on X, whence the equivalence classes determined by R_F, of the form

$$(27) \qquad R_F(x) = [x]_{R_F} = \{ y \in X : F(x) = F(y) \},$$

partition X. F is a constant mapping on each R_F-equivalence class. R_F is called the *equivalence relation on X induced by F*, and F is called a *generator* of this equivalence relation. All these are exactly analogous to the situation when a mapping $f : X \to W$ induces an equivalence relation R_f on X (*cf. ML*: 2.19). Note, however, that for a mapping $f : X \to W$, $\text{dom}(f) = \text{cor}(f) = X$, so $\text{dom}(f) \sim \text{cor}(f) = \varnothing$; but for a general set-valued mapping $F : X \dashv\subset W$, the R_F-equivalence class

(28) $F^{+1}(\varnothing) = \{x \in X : F(x) = \varnothing\} = X \sim \text{cor}(F) = \text{dom}(F) \sim \text{cor}(F)$

may not be empty.

The equivalence relation induced on a set X by a constant mapping is the universal relation $U = X \times X$ (with only one single partition block which is all of X; cf. ML:1.11). The equivalence relation induced on a set X by an injective mapping F includes the R_F-equivalence class (28) and partitions $\text{cor}(F)$ into singleton sets. Thus, if an injective mapping F has $F^{+1}(\varnothing) = X \sim \text{cor}(F)$ $= \varnothing$ (or when $F^{+1}(\varnothing)$ is a singleton set), the equivalence relation R_F induced on X is the equality relation I (with each partition block of X a singleton set).

3.19 Idempotence   When the set-valued mapping $Q : X \multimap X$ is an equivalence relation, the equivalence relation R_Q on X induced by Q, as in (27), has equivalence classes

(29) $$R_Q(x) = [x]_{R_Q} = \{y \in X : Q(x) = Q(y)\}.$$

Since Q is reflexive, $y \in Q(y)$, so with $Q(x) = Q(y)$ in (29) one has $y \in Q(x)$, whence $(x, y) \in Q$. Thus

(30) $$R_Q(x) = [x]_{R_Q} = \{y \in X : (x, y) \in Q\}.$$

But this is precisely the definition of the equivalence classes $[x]_Q$ of Q themselves, as in (25). In other words, one has

(31) $$R_Q = Q;$$

the equivalence relation induced on the domain by an equivalence-relation-as-a-set-valued mapping is the equivalence relation itself.

In particular, for a general set-valued mapping $F : X \multimap W$, the equivalence relation R_F on X induced by F is such that

(32) $$R_{R_F} = R_F;$$

that is, R_{R_F} and R_F are the same equivalence relation on X, and R_{R_F} partitions X the same way as R_F. Succinctly,

(33) $R_{R.} = R_.,$

and one may conclude that the operator $R_.$ is 'idempotent'.

The idempotence allusion of $R_.$ is more easily seen if (33) is represented symbolically with a relative product (i.e. sequential composition) as

(34) $R_. \circ R_. = R_..$

Monoids

3.20 The Monoid Rel(X, X)   A *monoid* $\langle M, * \rangle$ is an algebraic structure of a set M equipped with an associative binary operation $*$ and an identity element $e \in M$; equivalently, a monoid is a category with one object. Indeed, for any category **C** and any **C**-object X, the hom-set $\mathbf{C}(X, X)$ is a monoid (with the binary operation the composition of **C**-morphisms and the identity 1_X).

The hom-set $\mathbf{Rel}(X, X) = \mathbb{P}(X \times X)$ with the relative product (sequential composition) $\circ$ as the binary operation is therefore a *monoid*, with the identity set-valued mapping $1_X : X \multimap X$ as the monoid identity. One has, naturally, the property that for each $R : X \multimap X$, $R \circ 1_X = 1_X \circ R = R$.

It is worth noting that $\mathbf{Rel}(X, X)$ has two other set-valued mappings with distinguished properties. The *constant empty-set-valued mapping* $\varnothing : X \multimap X$ (Section 2.18) is such that for each $R : X \multimap X$, $R \circ \varnothing = \varnothing \circ R = \varnothing$. At the other extreme, the *universal relation* $U = X \times X$ is such that for each $R : X \multimap X$ with $R \neq \varnothing$, $R \circ U = U \circ R = U$.

It is important to note that $\mathbf{Rel}(X, X)$ is not a group. Although for each $R : X \multimap X$ there exists the inverse set-valued mapping $R^{-1} : X \multimap X$ (and that the inverse functor $Inv : \mathbf{Rel} \to \mathbf{Rel}$ defines the **Rel**-morphism assignment $Inv(R) = R^{-1}$; *cf.* Section 3.12,), the somewhat misnamed 'inverse' R^{-1} is not the inverse group element of R. Recall (*cf.* Sections 2.24 and 2.25) that one does not necessarily have $R^{-1} \circ R = 1_X$ and $R \circ R^{-1} = 1_X$.

3.21 Power Submonoid   In any monoid $\langle M, * \rangle$, one may define nonnegative integer powers of an element $x \in M$:

(35) $\begin{cases} x^0 = e \\ x^i = x * x^{i-1} \quad \text{for } i = 1, 2, 3, \ldots \end{cases}$

One easily verifies the rule of powers

(36) $$x^{i+j} = x^i * x^j$$

as well as the commutativity

(37) $$x^i * x^j = x^j * x^i.$$

So $\left\langle \{x^i : i \in \mathbb{N}_0\}, * \right\rangle$ is a *commutative submonoid* of $\langle M, * \rangle$.

In the monoid $\mathbf{Rel}(X, X)$, for a set-valued mapping $R : X \multimap X$, one may iteratively define, for $x \in X$,

(38) $$\begin{cases} R^0(x) = 1_X(x) = \{x\} \\ R^i(x) = R \circ R^{i-1}(x) \quad \text{for } i = 1, 2, 3, \dots \end{cases}$$

whence generating the commutative monoid $\left\langle \{R^i : i \in \mathbb{N}_0\}, \circ \right\rangle$. Beginning with the inverse set-valued mapping $R^{-1} : X \multimap X$, the iteration (38), with the further definition of

(39) $$R^{-i} = \left(R^{-1}\right)^i,$$

generates the commutative monoid $\left\langle \{R^{-i} : i \in \mathbb{N}_0\}, \circ \right\rangle$.

3.22 Example   Within each of the two commutative monoids $\left\langle \{R^i : i \in \mathbb{N}_0\}, \circ \right\rangle$ and $\left\langle \{R^{-i} : i \in \mathbb{N}_0\}, \circ \right\rangle$, sequential composition is commutative and satisfies the power rule; that is, if the indices i and j are both nonnegative or both nonpositive, then

(40) $$R^i \circ R^j = R^j \circ R^i = R^{i+j}.$$

But the equality (40) does not cross the zero boundary: if $i > 0$ and $j < 0$, then (40) is not necessarily satisfied. So the two submonoids of $\mathbf{Rel}(X, X)$ cannot be 'merged' into a 'subgroup' $\left\langle \{R^i : i \in \mathbb{Z}\}, \circ \right\rangle$.

Consider the simple example $R : \{1, 2\} \multimap \{1, 2\}$ with $R(1) = R(2) = \{1\}$; then $R^{-1}(1) = \{1, 2\}$ and $R^{-1}(2) = \varnothing$. One may calculate

$$(41) \quad \begin{cases} R^{-1} \circ R(1) = R^{-1}(\{1\}) = \{1,2\} \\ R^{-1} \circ R(2) = R^{-1}(\{1\}) = \{1,2\} \end{cases}.$$

On the other hand,

$$(42) \quad \begin{cases} R \circ R^{-1}(1) = R(\{1,2\}) = \{1\} \\ R \circ R^{-1}(2) = R^{-1}(\varnothing) = \varnothing \end{cases}.$$

Thus one sees that $R^{-1} \circ R \neq R \circ R^{-1}$, and neither composite is the identity $R^0 = 1_X$.

Closures

3.23 Operation Closure   An *n-ary operation* λ on a (nonempty) *set* S is a (single-valued) mapping $\lambda : S^n \to S$. The number n, called the *arity* of the operation, is usually a (finite) nonnegative integer, but may be naturally extended to be (countably or uncountably) infinite. A *nullary operation* (when $n = 0$), $\lambda : 1 \to S$, is the postulate of the existence of a special element $\lambda_0 \in S$. A *unary operation* (when $n = 1$), $\lambda : S \to S$, is a standard mapping that is a single-valued relation on S. One most often encounters *binary operations* $\lambda : S \times S \to S$, when $n = 2$.

A subset $A \subset S$ is *closed* under the operation λ (whence A has *closure* under the operation) if the result of that operation on elements of A is also included in A; that is, for all $x_1, x_2, ..., x_n \in A$, $\lambda(x_1, x_2, ..., x_n) \in A$, whence $\lambda(A^n) \subset A$. When $A \subset S$ is closed under λ, the operation $\lambda : S^n \to S$ when restricted to A defines the *n-ary operation* $\lambda : A^n \to A$ on A.

For example, the set $\mathbb{N}$ of natural numbers is closed under the binary operation $+ : \mathbb{R} \times \mathbb{R} \to \mathbb{R}$ of addition, since if $a, b \in \mathbb{N}$ then $+(a,b) = a+b \in \mathbb{N}$, so $+$ is also an operation on $\mathbb{N}$. But the subset $\mathbb{N} \subset \mathbb{R}$ is not closed under subtraction, since if $b \geq a$ then $a - b \notin \mathbb{N}$; one needs to expand $\mathbb{N}$ to the set $\mathbb{Z}$ for closure under subtraction. For another example, the set $\mathbb{Q} \sim \{0\}$ of nonzero rational numbers is closed under the unary operation $(\bullet)^{-1}$ of reciprocal, since if $a \in \mathbb{Q}$ and $a \neq 0$, then $1/a \in \mathbb{Q} \sim \{0\}$, but $\mathbb{N}$ is not closed under reciprocal.

Now consider the lattice $\langle \mathbb{P}(X \times X), \cup, \cap \rangle$ (*cf. ML*: 2.1) of all relations on X. I shall examine whether certain special subsets of $\mathbb{P}(X \times X)$ are closed under the meet $\wedge = \cap$ and the join $\vee = \cup$ operations.

Let $R \subset P(X \times X)$ be the set of all reflexive relations on X. The intersection of two reflexive relations is itself reflexive, so is their union. Thus the set R is closed under binary intersection and binary union. One may trivially extend the intersection to be taken over n reflexive relations for any finite n, and one sees that the set R is closed under the n-nary operation of intersection. Indeed, the extension may be taken to arbitrary (all finite and infinite, countable and uncountable) collections, so the set R is closed under arbitrary intersections. Note that this includes the trivial unary operation of 'intersection of one set' (i.e. the identity operation on $P(X \times X)$) and the nullary operation of intersection over the empty collection (which yields $\bigcap_\varnothing = X \times X$; cf. ML: 1.28). The closure of R under arbitrary intersections is, therefore, only true if the universal relation $U = X \times X$ is reflexive (which it is; cf. Section 3.16). On the other hand, the set R is closed under unions only over any nonempty family (since $\bigcup_\varnothing = \varnothing$ is not reflexive).

The set $S \subset P(X \times X)$ of all symmetric relations is closed under arbitrary intersections and arbitrary unions.

The set $T \subset P(X \times X)$ of all transitive relations on X is likewise closed under arbitrary intersections. Since the set $QX \subset P(X \times X)$ of equivalence relations on X is $QX = R \cap S \cap T$, it is also closed under arbitrary intersections (whence the lattice QX is closed under the *meet* operation $\wedge = \cap$ over arbitrary collections; cf. ML: 2.16). The union of two transitive relations is, however, not necessarily transitive (so T is not closed under the binary operation $\cup$). This is because, for $R_1, R_2 \in T$, $x R_1 y$ and $y R_2 z$, for example, do not automatically imply $x(R_1 \cup R_2)z$. This is an example that shows to expand a subset $A \subset S$ to a closed set under an operation $\lambda : S^n \to S$, it is not always a simple matter of taking the union $A \cup \lambda(A^n)$; that is, $A \cup \lambda(A^n)$ is not necessarily closed under λ. We have encountered this situation in ML: 2.17—the *join* of two equivalence relations in the lattice QX of equivalence relations on X has to be expanded to the *transitive closure* of their union, not simply the union itself. (I shall revisit transitive closure presently.)

The set $A \subset P(X \times X)$ of all antisymmetric relations is closed under intersections over any nonempty family (since $\bigcap_\varnothing = X \times X$ is not antisymmetric). The union of two antisymmetric relations is not necessarily antisymmetric (so A is not closed under the binary operation $\cup$), as the following simple example shows. Let $X = \{1, 2\}$, $R_1 = \{(1,2)\}$, and $R_2 = \{(2,1)\}$. Then $R_1, R_2 \in A$, and $R_1 \cup R_2 = \{(1,2),(2,1)\}$. Since $(1,2),(2,1) \in R_1 \cup R_2$, antisymmetry would require the impossible $1 = 2$. Indeed, the condition (a) of antisymmetry $(cf.$ Definition 3.14(a) and Lemma 3.16(a)) is a restrictive one, the contrapositive of the defining statement being

$$(43) \qquad\qquad x \neq y \;\Rightarrow\; \neg\left(xRy \text{ and } yRx \right).$$

Statements (43) and (a) therefore dictate what cannot be members of an antisymmetric relation. (Contrariwise, statements (r), (s), and (t) stipulate what, respectively, a reflexive, symmetric, and transitive relation needs to have as members.) If there are two different elements $x, y \in R$ with xRy and yRx (whence $R \notin \mathcal{A}$), there is no way to 'expand' R to some larger relation in $\mathcal{A}$. Viewed otherwise, as per Lemma 3.16(a), the condition for antisymmetry is that of an upper bound: R is antisymmetric iff $R \cap R^{-1} \subset I$; so if $R \cap R^{-1}$ exceeds its containment by the upper bound I , no 'expansion' will make it smaller.

3.24 Closure Operator   A *closure operator* on a set X is a mapping $c : PX \to PX$ of power sets such that, for all $A, B \in PX$:

i.   *Extensive*: $A \subset c\left(A \right)$

ii. *Idempotent*: $c\left(c\left(A \right) \right) = c\left(A \right)$

iii. *Monotone (increasing)*: $A \subset B \;\Rightarrow\; c\left(A \right) \subset c\left(B \right)$

The set $c\left(A \right)$ is called the *closure* of A , and $A \in PX$ is called *closed* (and that it is a *closed set*) if $A = c\left(A \right)$. Because of the idempotence ii, a subset of X is closed iff it is the closure of some subset of X ; that is, the closed subsets of X are precisely those of the form $c\left(A \right)$. The defining properties of a closure operators imply that the closure $c\left(A \right)$ of a set A is the smallest (in the poset $\langle PX, \subset \rangle$) closed set containing A .

The collection C of all closed set defined by the *closure operator* $c : PX \to PX$ *on* X is thus $\mathsf{C} = c\left(PX \right)$, and $\mathsf{C} = c\left(\mathsf{C} \right) \subset PX$, so *a fortiori* $c\left(\mathsf{C} \right) \subset \mathsf{C}$. The set $\mathsf{C} \subset PX$ is therefore *closed* under the *unary operation* $c : PX \to PX$ *on* PX (in the sense of the previous section), hence the linkage between the two nuanced concepts of 'closure'. Note the distinction in the customary terminology: the mapping of power sets, $c : PX \to PX$, when considered as a closure operator is 'on X' and when considered as a closure-defining unary operation is 'on PX'.

The usage of 'closure' originated, of course, in topology, in which the closure $\overline{A}$ of a set A in a topological space is the union of A and the set of its accumulation points (and equivalently the intersection of the members of the family of all closed sets—complements of open sets that define the topology—containing A). The map $A \mapsto \overline{A}$ is a closure operator on the topological space.

3.25 Closure of a Relation   The notion of a closure (in the closure operator sense) when specialized to relations on a set X appears thus. For an arbitrary relation $R \in P(X \times X)$ and an arbitrary property p on $P(X \times X)$, the p *closure of* R , if it exists, is the smallest relation $Q \in P(X \times X)$ (in the poset $\langle P(X \times X), \subset \rangle$) that contains R and for which property p holds; that is, the relation

$$(44) \qquad \inf \{ Q \in P(X \times X) : R \subset Q \text{ and } p(Q) \}$$

(*cf. ML*: 1.27). Note the qualifier 'if it exists' in the previous sentence: for arbitrary p and R , the p closure of R need not exist. If sets with property p are closed under arbitrary intersections, however, then the p *closure of* R does exist and may be directly defined as the intersection of all sets containing R and with property p (since $\langle P(X \times X), \cup, \cap \rangle$ is a *complete lattice* in which the meet operation is $Q_1 \wedge Q_2 = \inf \{ Q_1, Q_2 \} = Q_1 \cap Q_2$; *cf. ML*: 2.10). (As usual, 'arbitrary intersections' include the intersection over the empty family, so the universal relation $U = X \times X$ must also have the property p .) The assignment $R \mapsto Q$, which maps a relation $R \in P(X \times X)$ to its p closure Q , is a closure operator on $P(X \times X)$.

Thus, in particular, the *reflexive closure, symmetric closure*, and *transitive closure* of a relation R on a set X (i.e. set-valued mapping $R : X \multimap X$) all exist and are, respectively, the smallest reflexive, symmetric, and transitive relation on X that contains R and equivalently the intersection of all the reflexive, symmetric, and transitive relations on X that contain R . They have, however, alternate characterizations that are more constructive.

3.26 Lemma   *The reflexive closure* $R^=$ *of the set-valued mapping* $R : X \multimap X$ *is*

$$(45) \qquad\qquad\qquad R^= = I \cup R.$$

The reflexive closure $R^= : X \multimap X$ of R is thus the union of the identity mapping $1_X = I : X \multimap X$ with R . The union mapping of two set-valued mappings is defined pointwise as in Definition 2.28.i, so for $x \in X$,

$$(46) \qquad\qquad R^=(x) = I(x) \cup R(x) = \{x\} \cup R(x).$$

3.27 Lemma   *The symmetric closure $R^{\pm}$ of the set-valued mapping $R: X \multimap X$ is*

(47) $$R^{\pm} = R \cup R^{-1}.$$

The symmetric closure $R^{\pm}: X \multimap X$ of R is the union of R with its inverse mapping $R^{-1}: X \multimap X$; defined pointwise, this is, for $x \in X$,

(48) $$R^{\pm}(x) = R(x) \cup R^{-1}(x).$$

One may see that $R^{\pm}$ is indeed a symmetric relation in the equivalences

(47)
$$\begin{aligned}
&y \in R^{\pm}(x) \\
\Leftrightarrow\ &y \in R(x) \cup R^{-1}(x) \\
\Leftrightarrow\ &y \in R(x) \text{ or } y \in R^{-1}(x) \\
\Leftrightarrow\ &xRy \text{ or } yRx \\
\Leftrightarrow\ &x \in R^{-1}(y) \text{ or } x \in R(y) \\
\Leftrightarrow\ &x \in R(y) \cup R^{-1}(y) \\
\Leftrightarrow\ &x \in R^{\pm}(y)
\end{aligned}$$

Recall (Section 3.22) that the two commutative power submonoids $\langle \{R^i : i \in \mathbb{N}_0\}, \circ \rangle$ and $\langle \{R^{-i} : i \in \mathbb{N}_0\}, \circ \rangle$ of $\mathbf{Rel}(X, X)$ cannot be 'merged' into a 'cyclic subgroup' $\langle \{R^i : i \in \mathbb{Z}\}, \circ \rangle$. With the union (47), one sees that one may consider the power submonoid $\langle \{(R^{\pm})^i : i \in \mathbb{N}_0\}, \circ \rangle$ generated by the symmetric closure $R^{\pm}$ in its stead.

3.28 Lemma   *The transitive closure $\vec{R}$ of the set-valued mapping $R: X \multimap X$ is*

(48) $$\vec{R} = R \cup R^2 \cup R^3 \cup \cdots = \bigcup_{i \in \mathbb{N}} R^i.$$

(The powers R^i are as defined in Section 3.21.)   Pointwise, the set-valued mapping $\vec{R}: X \multimap X$ is defined by, for $x \in X$,

(49) $$\vec{R}(x) = R(x) \cup R^2(x) \cup R^3(x) \cup \cdots.$$

$y \in \vec{R}(x)$ iff $y \in R^i(x)$ for some $i \in \mathbb{N}$. This means there is a finite sequence of elements $x_1, x_2, ..., x_{i-1} \in X$ such that

$$(50) \qquad\qquad x R x_1, x_1 R x_2, ..., x_{i-1} R y$$

(*cf.* ML: 2.17, in which I defined the join $R_1 \vee R_2$ of two equivalence relations R_1 and R_2 as the transitive closure $\left(R_1 \cup R_2\right)^{\rightarrow}$ of their union $R_1 \cup R_2$ —as I explained in Section 3.23 above, R and S are closed under binary union, but not T).

    I leave it as an exercise to the reader to verify that each of the three closure assignments—reflexive closure $R \mapsto R^{=}$, symmetric closure $R \mapsto R^{\pm}$, and transitive closure $R \mapsto \vec{R}$—satisfies the three properties in Section 3.24 (extension, idempotence, and monotonicity), whence each defines a closure operator $c : P(X \times X) \rightarrow P(X \times X)$.

3.29 Commutation   I also leave it as an exercise for the reader to show that the three closure operators on $P(X \times X)$—reflexive closure $(\bullet)^{=}$, symmetric closure $(\bullet)^{\pm}$, and transitive closure $(\bullet)^{\rightarrow}$—commute pairwise with one another. Explicitly, for all set-valued mappings $R : X \multimap X$,

reflexive symmetric closure = symmetric reflexive closure:

$$(51) \qquad\qquad \left(R^{\pm}\right)^{=} = \left(R^{=}\right)^{\pm};$$

reflexive transitive closure = transitive reflexive closure:

$$(52) \qquad\qquad \left(\vec{R}\right)^{=} = \left(R^{=}\right)^{\rightarrow};$$

symmetric transitive closure = transitive symmetric closure:

$$(53) \qquad\qquad \left(\vec{R}\right)^{\pm} = \left(R^{\pm}\right)^{\rightarrow}.$$

The reflexive symmetric transitive closure of R (with the three closures applied in any order) is, naturally, the *equivalence closure* $R^{\equiv}$ of R, that is, the smallest equivalence relation that contains R.

3.30 Eightfold Partition   Each of the three closure operators may act on a set-valued mapping $R: X \multimap X$ zero or one times, yielding eight possibly distinct members of $P(X \times X)$:

$$(54) \qquad\qquad R, \; R^{=}, \; R^{\pm}, \; \vec{R}, \; \left(R^{\pm}\right)^{=}, \; \left(\vec{R}\right)^{=}, \; \left(\vec{R}\right)^{\pm}, \; R^{=}.$$

Because of the idempotence and commutativity of these closure operators, their repeated applications to a set-valued mapping $R: X \multimap X$ will not yield additional members of $P(X \times X)$. For example, $\left(\left(\left(\vec{R}\right)^{=}\right)^{\rightarrow}\right)^{=} = \left(\left(\left(\vec{R}\right)^{\rightarrow}\right)^{=}\right)^{=} = \left(\vec{R}\right)^{=}$.

Note that I wrote the list (54) yields 'eight possibly distinct members of $P(X \times X)$'. It is entirely possible that some of the subsets on the list coincide (i.e. the closure of one property also adds another). For example, let $X = \{1,2,3\}$, and let the set-valued mapping $R: X \multimap X$ be defined by $R(1) = \{3\}$, $R(2) = \{1,3\}$, and $R(3) = \{2\}$.   One may verify that the transitive closure of R is $\vec{R} = X \times X = U$, which is also reflexive and symmetric; that is, $\vec{R} = R^{=}$.

The three subsets $\mathbf{R}, \mathbf{S}, \mathbf{T} \subset P(X \times X)$ of, respectively, reflexive, symmetric, and transitive relations on X (*cf.* Section 3.23) partition their Venn diagram into eight distinct regions:

(55)

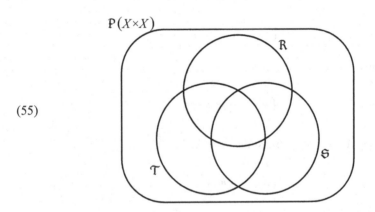

For $|X| > 2$, each region is nonempty; that is, for example, there are relations that are reflexive and symmetric but not transitive.

Let me give you an illustrative example, a relation R for which the eight combinations of the three closures in (54) just happen to occupy the eight regions.

Let $X = \{1,2,3,4\}$, and let the set-valued mapping $R : X \multimap X$ be defined by
$R(1) = \{2\}$, $R(3) = \{1\}$, and $R(2) = R(4) = \varnothing$. Then, since $1 \notin R(1)$, R is not
reflexive; since $2 \in R(1)$ but $1 \notin R(2)$, R is not symmetric; and since $3R1$ and
$1R2$ but not $3R2$, R is not transitive. Thus

$$(56) \qquad\qquad R = \{(1,2),(3,1)\} \in \mathbf{R}^c \cap \mathbf{S}^c \cap \mathbf{T}^c.$$

For this R, each of the three closures just adds that single property but not the
other two. The seven closures of R are distinct from one another:

$$
\begin{aligned}
R^= &= \{(1,1),(1,2),(2,2),(3,1),(3,3),(4,4)\} \in \mathbf{R} \cap \mathbf{S}^c \cap \mathbf{T}^c \\
R^\pm &= \{(1,2),(1,3),(2,1),(3,1)\} \in \mathbf{R}^c \cap \mathbf{S} \cap \mathbf{T}^c \\
\vec{R} &= \{(1,2),(3,1),(3,2)\} \in \mathbf{R}^c \cap \mathbf{S}^c \cap \mathbf{T} \\
(57) \quad (R^\pm)^= &= \{(1,1),(1,2),(1,3),(2,1),(2,2),(3,1),(3,3),(4,4)\} \in \mathbf{R} \cap \mathbf{S} \cap \mathbf{T}^c \\
(\vec{R})^= &= \{(1,1),(1,2),(2,2),(3,1),(3,2),(3,3),(4,4)\} \in \mathbf{R} \cap \mathbf{S}^c \cap \mathbf{T} \\
(\vec{R})^\pm &= \{(1,1),(1,2),(1,3),(2,1),(2,2),(2,3),(3,1),(3,2),(3,3)\} \in \mathbf{R}^c \cap \mathbf{S} \cap \mathbf{T} \\
R^* &= \{(1,1),(1,2),(1,3),(2,1),(2,2),(2,3),(3,1),(3,2),(3,3),(4,4)\} \in \mathbf{R} \cap \mathbf{S} \cap \mathbf{T}
\end{aligned}
$$

Partial Orders

3.31 Partial Order   Let $R : X \multimap X$ be a set-valued mapping that is (r)
reflexive, (a) antisymmetric, and (t) transitive. If one writes

$$(58) \qquad\qquad y \le x \text{ iff } y \in R(x),$$

then one may verify that R defines a partial order $\le$ on the set X (*cf. ML*: 1.20).
(Note that $x \in R(x)$ so $\operatorname{ran}(R) = \operatorname{cor}(R) = X$.)

   Conversely, given a poset $\langle X, \le \rangle$, the mapping $R : X \to PX$ defined, for
$x \in X$, by

$$(59) \qquad\qquad R(x) = \{y \in X : y \le x\},$$

embeds $\langle X, \le \rangle$ as a suborder of $\langle PX, \subset \rangle$:

$$(60) \qquad\qquad \langle X, \le \rangle \cong \langle R(X), \subset \rangle$$

(where $R(X) = \{R(x) : x \in X\} \subset PX$; cf. ML: 1.24).

3.32 Dual Order   For the partial order (58) defined by a set-valued mapping, since $y \leq x$ iff $y \in R(x)$ iff $x \in R^{-1}(y)$, one sees that, as expected, the *dual* (or *converse*) partial order $\geq$ on X is defined by the inverse of R:

(61) $y \geq x$ iff $y \in R^{-1}(x)$.

(In the notation of *ML*: 1.25, this says $\breve{R} = R^{-1}$.)

3.33 Maximal Element   Let $\leq$ be a partial order on a set X and let $A \subset X$. Recall (*ML*: 1.29) that an element $x \in A$ is *maximal* if whenever $y \in A$ and $x \leq y$, one has $x = y$. Succinctly, $x \in A$ is maximal if $x < y$ for no $y \in A$. When the partial order $\leq$ is defined by a set-valued mapping R, therefore, $x \in A$ is maximal if for every $y \in A \sim \{x\}$, $x \notin R(y)$.

Graphs

3.34 Directed Graph   Let $R : X \multimap X$ be a set-valued mapping.   Recall (Definition 2.2) that R in its relational form may be called its *graph*:

(62) $R = \{(x, y) \in X \times X : y \in R(x)\} = \{(x, y) \in X \times X : (x, y) \in R\} \subset X \times X$.

In addition to its standard-mapping sense of a diagram or a picture of a relation among variables, this definition of 'graph' of a set-valued mapping also has linkages to its topological sense of a collection of vertices connected by (directed) edges.

 If two elements $x, y \in X$ are such that $y \in R(x)$, one says that x *is linked to* y *by the relation* R. In a topological network that is a directed graph (*digraph*), this may be represented by joining the two points by a *directed edge from vertex* x *to vertex* y (cf. *ML*: 6.5). The set $R(x)$ thus contains all the vertices in X that terminate an arrow (i.e. a directed edge) initiating from x. Stated otherwise, $R(x)$ contains all the vertices in X *reachable* from x after travelling on exactly one directed edge. If $x \in R(x)$ (the reflexivity, of course, may or may not happen), then there is a *self-loop* from x to itself. In the digraph R, for two vertices $x, y \in X$, there can be at most one directed edge from vertex x to vertex y (one edge if $y \in R(x)$ and zero edges if $y \notin R(x)$, by definition). There may, however, be a reverse edge: there is either one or zero directed edges from vertex y to vertex x (depending on whether $x \in R(y)$ or equivalently $y \in R^{-1}(x)$).

Conversely, given a digraph R with at most one directed edge from one vertex to another, the same association of a directed edge from vertex x to vertex y with $y \in R(x)$ defines a relation R on the set X of vertices. Thus there is a one-to-one correspondence between relations and this collection of digraphs.

3.35 Injective Digraph   If R is *injective* (Definition 2.6), then its defining property $x \neq y \implies R(x) \cap R(y) = \varnothing$ means that arrows initiating from different vertices cannot converge on the same terminating vertex. The set $R^{-1}(x)$ contains all the vertices in X initiating an arrow that terminates on x. Thus for an injective digraph R, each $R^{-1}(x)$ is either empty or a singleton set.

3.36 Reachability   In the digraph defined by a set-valued mapping $R : X \multimap X$, the set $R(x)$ contains all the vertices in X *reachable* from the vertex x after travelling on exactly one directed edge. Vertices in X reachable from vertex x after travelling on two connected edges (oriented in the appropriate directions) in the digraph R are contained in the set $R^2(x)$, those after three connected edges are in $R^3(x)$, etc. Any vertex is by default reachable from itself on zero connected edges; one may represent this as $x \in \{x\} = I(x)$. Note that this is different from the existence of a self-loop, on which a vertex is reachable from itself after travelling on one edge; this happens iff $x \in R(x)$.

All the vertices in the digraph R that are reachable from a vertex $x \in X$ by a trace of directed edges are thus members of the set

(63) $$\hat{R}(x) = \{x\} \cup R(x) \cup R^2(x) \cup R^3(x) \cup \cdots = I(x) \cup \vec{R}(x),$$

that is, the set defined by the reflexive closure of the transitive closure of R (*cf.* Lemmata 3.26 and 3.28). As a set-valued mapping, one may, naturally, call $\hat{R} : X \multimap X$, hence defined by

(64) $$\hat{R} = I \cup \vec{R} = \left(\vec{R}\right)^=,$$

the *reflexive transitive closure* (= transitive reflexive closure; *cf.* Section 3.29) of R.

The 'reachability' relation on X for a set-valued mapping $R : X \multimap X$, defined for $x, y \in X$ by $y \in \hat{R}(x)$, whence there is a path in the digraph R from x to y, is evidently, then, reflexive and transitive. It is therefore a preorder (*cf.*

Section 0.22); indeed, $\hat{R}$ is the *preorder closure* of R, it being the smallest preorder containing R. Reachability is, however, not symmetric: ' y is reachable from x' does not necessarily imply ' x is reachable from y ' on a digraph.

The equality

(65)
$$
\begin{aligned}
R \circ \hat{R}(x) &= R(\{x\} \cup R(x) \cup R^2(x) \cup R^3(x) \cup \cdots) \\
&= R(x) \cup R^2(x) \cup R^3(x) \cup \cdots
\end{aligned}
$$

shows that

(66)
$$
R \circ \hat{R} = R \circ \left(I \cup \vec{R} \right) = \vec{R}.
$$

If indeed

(67)
$$
x \in R \circ \hat{R}(x) = \vec{R}(x),
$$

then vertex x is reachable from itself after travelling on one or more edges, so the vertex x is involved in a (directed) *cycle* (which includes the self-loop case of $x \in R(x)$), whence the digraph R contains a cycle.

3.37 Undirected Graph   Occasionally it may be useful to consider the graph not in the directed sense (so that one may travel on an edge in both directions). One sees that the set $R(x) \cup R^{-1}(x)$ contains all the vertices in the digraph R that are directly connected with x (i.e. all the vertices connected by a single directed edge to or from x). Indeed, given a set-valued mapping $R: X \multimap X$, one may define its associated *undirected graph* by its *symmetric closure* (*cf.* Lemma 3.27) $R^{\pm}: X \multimap X$: for $x \in X$,

(68)
$$
R^{\pm}(x) = R(x) \cup R^{-1}(x).
$$

In the undirected graph $R^{\pm} = R \cup R^{-1}$, the directed edge from x to y and the directed edge from y to x, if either one exists, merge into an undirected edge joining x and y (the merger neatly taken care of by the union operation); thus in $R^{\pm}$ there are no *multiple edges* (which are two or more edges joining the same two vertices in a graph; *cf.* ML: 6.3). If one further disallows self-loops (by imposing on all vertices $x \in X$ the condition $x \notin R(x)$), then the undirected graph $R^{\pm}$ is a *simple graph* (which is by definition a graph that contains no self-loops and no multiple edges; *cf.* ML: 6.3).

3.38 Path   A *directed path* (also called a *dipath*) is a sequence of vertices $\{x_0, x_1, x_2, ..., x_n\}$ in X such that

(69) $x_i \in R(x_{i-1})$ for $i = 1, 2, ..., n$.

On this path, one may trace n directed edges in sequence (in the proper direction of each edge) and travel from vertex x_0 to vertex x_n. An *undirected path* (or simply *path*) may be defined as a sequence of vertices $\{x_0, x_1, x_2, ..., x_n\}$ in X such that

(70) $x_i \in R^{\pm}(x_{i-1}) = R(x_{i-1}) \cup R^{-1}(x_{i-1})$ for $i = 1, 2, ..., n$.

One may still travel from vertex x_0 to vertex x_n on this undirected path, if one ignores the directions of the edges (*cf. ML*: 6.3). The *length* n of a path (directed or undirected) is the number of edges in it. Vacuously, by default, there is a dipath [/path] of zero directed [/undirected] edges (i.e. of length 0) from every vertex $x \in X$ to itself.

3.39 Definition   For a fixed element $x \in X$, the indexed family of sets $\{R^i(x)\}_{i \in \mathbb{N}_0}$ is called the *trajectory of x under R*.

3.40 Connectedness   In the undirected graph $R^{\pm}$, the collection of all the vertices that are reachable from a vertex $x \in X$ by a trace of edges (regardless of the directions in the digraph R) is the set $\hat{R}^{\pm}(x) \subset X$, which, incidentally, is the union of the member sets of the trajectory of x under $R^{\pm}$:

(71) $\hat{R}^{\pm}(x) = \bigcup_{i \in \mathbb{N}_0} \left(R^{\pm}\right)^i (x) = \bigcup_{i \in \mathbb{Z}} R^i(x)$

(*cf.* Definition 3.39). The set-valued mapping $\hat{R}^{\pm} : X \multimap X$ may be assembled as $R : X \multimap X$ sequentially operated on by three closure operators,

(72) $\hat{R}^{\pm} = \left(\left(R^{\pm}\right)^{\rightarrow}\right)^{=} = I \cup \left(R \cup R^{-1}\right) \cup \left(R \cup R^{-1}\right)^2 \cup \left(R \cup R^{-1}\right)^3 \cup \cdots,$

that is, the reflexive transitive symmetric closure of R, which is, of course, the equivalence closure $R^{=}$ of R (*cf.* Section 3.29). Indeed, the relation of reachability in the undirected graph $R^{\pm} : X \multimap X$, defined for $x, y \in X$ by

(73) $y \in \hat{R}^{\pm}(x),$

that there is a path from x to y is an equivalence relation on the set X of vertices. (Recall that reachability in a digraph is not symmetric.)

If $\hat{R}^{\pm}(x) = X$, then every vertex may be reached from vertex x on an undirected path. This, in fact, implies that there is an undirected path from any vertex to any other vertex (by travelling on two undirected paths with a 'transfer' at vertex x if necessary), that is, if $\hat{R}^{\pm}(x) = X$ for some $x \in X$, then the inclusion holds for all $x \in X$. A graph with this property, that there is a path from any vertex to any other vertex, is called *connected* (*cf. ML*: 6.3).

Even if the whole graph may not be connected, the equivalence relation (73) partitions X into equivalence classes called the *connected components*. Vertices in each connected component are reachable from one another on undirected paths, while vertices from different connected components are not. Each connected component is a maximal connected subgraph of the undirected graph $R^{\pm}$.

4
Censusing Independence

According to the number that you offer, so you shall do with
each and every one.

— *Numeri* 15:12

In this chapter, let X be a set and F be a relation on X, i.e. a set-valued
mapping $F : X \multimap X$.

Mitto: **Isolation and Independence**

4.1 The Symmetric Closure Since F has the same set X as its domain and
codomain, so does F^{-1}, whence

(1) $$\operatorname{dom}(F) = \operatorname{cod}(F) = \operatorname{dom}(F^{-1}) = \operatorname{cod}(F^{-1}) = X.$$

Also, the ranges and coranges are related thus:

(2) $$\operatorname{ran}(F) = \operatorname{cor}(F^{-1}) = F(X)$$

and

(3) $$\operatorname{cor}(F) = \operatorname{ran}(F^{-1}) = F^{-1}(X),$$

but, of course, the two subsets (2) and (3) of X are in general different.

The inherent domain-codomain symmetry allows the definition of a
symmetric relation $F^{\pm}$ on X, the *symmetric closure* of F, that is, the bilateral
'undirected graph' mapping $F^{\pm} = F \cup F^{-1} : X \multimap X$:

A.H. Louie, *The Reflection of Life*, IFSR International Series on Systems Science
and Engineering 29, DOI 10.1007/978-1-4614-6928-5_4,
© Springer Science+Business Media New York 2013

(4) $$F^{\pm}(x) = F(x) \cup F^{-1}(x)$$

(*cf.* Lemma 3.27 and Section 3.37). Although one may define the union (4) of the two *sets* $F(x)$ and $F^{-1}(x)$ even when F has different sets for domain and codomain, the *union set-valued mapping* $F^{\pm} = F \cup F^{-1}$ (*cf.* Definition 2.28.i) is only defined when $\operatorname{dom}(F) = \operatorname{cod}(F)$. In any case, as explained in Section 3.3, when two sets are too different, their union has little consequence. But with both $F(x) \subset X$ and $F^{-1}(x) \subset X$, the union $F^{\pm}(x) = F(x) \cup F^{-1}(x) \subset X$ is their *join* (i.e. *supremum*) in the lattice $\langle PX, \cup, \cap \rangle$ (*ML*: 2.12). In what follows we shall see frequent apparitions of $F^{\pm} = F \cup F^{-1}$.

4.2 Definition   An element $x \in X$ is *isolated* (under the set-valued mapping $F : X \multimap X$ or, equivalently, under the relation F on X) if both sets $F(x)$ and $F^{-1}(x)$ are empty.

If $F(x) = \varnothing$, then $x \in F^{+1}(\varnothing)$ (*cf.* Section 2.16); so the collection of all isolated points is a subset of $F^{+1}(\varnothing)$.

   Since $F(x) = \varnothing$ iff $x \notin \operatorname{cor}(F)$, and $F^{-1}(x) = \varnothing$ iff $x \notin \operatorname{ran}(F)$, one sees that $x \in X$ is isolated iff $x \notin \operatorname{cor}(F) \cup \operatorname{ran}(F)$. On account of (2) and (3) (and the de Morgan law $A^c \cap B^c = (A \cup B)^c$), $x \in X$ is isolated iff

(5) $$x \notin F^{\pm}(X) = F(X) \cup F^{-1}(X).$$

If an element is isolated, then there are no edges attached to its corresponding vertex in the undirected graph $F^{\pm}$ (and vice versa); so each isolated element forms its own connected component (*cf.* Section 3.40).

4.3 Definition   Two elements $x, y \in X$ are *independent* (under the set-valued mapping $F : X \multimap X$, i.e. under the relation F on X) if $x \notin F(y)$ and $y \notin F(x)$.

Note that here again is where the equality $\operatorname{dom}(F) = \operatorname{cod}(F)$ inherent in F being a relation on X is crucial to the development of the concepts. The statements $x \in F(y)$, $x \notin F(y)$, etc., are significantly more interesting if the entities $x \in \operatorname{dom}(F)$, $F(y) \subset \operatorname{cod}(F)$, etc., belong to the same the lattice $\langle PX, \cup, \cap \rangle$, allowing for their possible relationships.

Since $y \notin F(x)$ iff $x \notin F^{-1}(y)$, the defining conditions for independence may equivalently be ' $x \notin F(y)$ and $x \notin F^{-1}(y)$ '; likewise, it may also be ' $y \notin F^{-1}(x)$ and $y \notin F(x)$ '. Stated otherwise, two elements $x, y \in X$ are independent iff

$$(6) \qquad\qquad x \notin F^{\pm}(y) = F(y) \cup F^{-1}(y),$$

or equivalently iff

$$(7) \qquad\qquad y \notin F^{\pm}(x) = F(x) \cup F^{-1}(x).$$

What is important is that the notion of independence is defined symmetrically: The unilateral declaration ' x is independent of y ' is not defined. Independence always implies mutuality, pairwise, 'each to the other'. It does not preclude, however, the selection $x = y$, when the two elements of X under consideration are the same. 'Two' elements $x, x \in X$ are independent, by definition, iff $x \notin F(x)$ (which is also equivalent to $x \notin F^{-1}(x)$), whence iff

$$(8) \qquad\qquad x \notin F^{\pm}(x) = F(x) \cup F^{-1}(x).$$

In this case one may say ' x is independent of itself'. Since reflexivity means, for all $x \in X$, $x \in F(x)$ (Definition 3.14(r)), no element can be independent of itself under a reflexive set-valued mapping.

For an isolated element $x \in X$, $F(x) = \varnothing$ implies that for all $y \in X$ $y \notin F(x)$, and $F^{-1}(x) = \varnothing$ implies that for all $y \in X$ $x \notin F(y)$. Thus, for any two elements of X, if at least one of them is isolated, then they are independent. If one considers that 'an isolated element $x \in X$ ' means

$$(9) \qquad\qquad F^{\pm}(x) = F(x) \cup F^{-1}(x) = \varnothing,$$

while 'two independent elements $x, y \in X$ ' means

$$(10) \qquad\qquad F^{\pm}(x) \cap \{y\} = \left[F(x) \cup F^{-1}(x) \right] \cap \{y\} = \varnothing,$$

the implication $(9) \Rightarrow (10)$ is even more evident.

4.4 Definition A subset Y of X is called an *independent set* (under the set-valued mapping $F : X \multimap X$, i.e. under the relation F on X) if every two elements of Y are independent.

One readily sees the

4.5 Lemma   *A subset Y of X is an independent set iff $Y \cap F(Y) = \varnothing$ iff $Y \cap F^{-1}(Y) = \varnothing$ (whence iff $Y \cap F^{\pm}(Y) = \varnothing$).*

Nuances of Independence

4.6 Self-Independence   Consider the simple example of a set-valued mapping $F : X \multimap X$ with $X = \{1,2\}$ and $F(1) = \{1\}$, $F(2) = \{2\}$. Then $1 \notin F(2)$ and $2 \notin F(1)$, so 1 and 2 are independent. But $X = \{1,2\}$ is not an independent set, because neither 1 nor 2 is independent of itself. Note that Definition 4.4 requires every two elements of an independent set to be independent, so it includes the 'self-independence' of all of its elements.

If $Y \subset X$ is independent, then all subsets of Y containing two or more elements clearly inherit the 'every two elements are independent' property, so they are all independent sets. All singleton subsets of Y, as explained in the previous paragraph, are also by definition independent. For the empty set $\varnothing$, the pairwise independence of its elements is vacuously true. Thus for an independent set Y, the property of independence is inherited by all its subsets, i.e. by every member of PY.

Henceforth in this chapter, I shall further assume that *all singleton sets are independent sets* under the set-valued mapping $F : X \multimap X$; i.e. each $x \in X$ is independent of itself, so one must have

(11) $x \notin F(x)$, or equivalently, $F(x) \subset X \sim \{x\}$.

As noted in Section 4.3 above, this requirement of elemental self-independence excludes all reflexive relations, so $I \cap F = \varnothing$. *A fortiori*, all subsets $Y \subset X$ with $|Y| = 0$ or 1 are independent sets under such F (where $|Y|$ denotes the cardinality, i.e. the number of elements, of the set Y; *cf.* Sections 0.9 and 0.10). The restriction (11) is also equivalent to each of the conditions in the following list: For all $x \in X$,

(12) $x \notin F^{-1}(x)$, $F^{-1}(x) \subset X \sim \{x\}$, $x \notin F^{\pm}(x)$, $F^{\pm}(x) \subset X \sim \{x\}$,

whence $I \cap F^{-1} = \varnothing$ and $I \cap F^{\pm} = \varnothing$.

For an element $x \in X$ not to be isolated means $F^{\pm}(x) \neq \varnothing$; with the added requirement $x \notin F^{\pm}(x)$, this means $F^{\pm}(x) \sim \{x\} \neq \varnothing$. Thus, for a 'non-isolated' element $x \in X$ (under our now-default set-valued mapping $F : X \multimap X$ with the property that for each $x \in X$, $x \notin F(x)$), there exists $y \in F^{\pm}(x)$ with $y \neq x$, i.e. there exists $y \in F^{\pm}(x) \sim \{x\}$.

4.7 Maximal Independent Set   The collection of all independent sets under a set-valued mapping $F : X \dashrightarrow X$ is a poset ordered by set inclusion $\subset$, i.e. a subset of the poset $\langle PX, \subset \rangle$. As such, one may consider the notion of a maximal element in this poset (*cf.* Section 3.33). Explicitly, a *maximal independent set* $Y \subset X$ is an independent set under the set-valued mapping $F : X \dashrightarrow X$, such that no element $x \in X \sim Y$ may be appended to it to make a larger independent set $Y \cup \{x\}$. In other words, when Y is a maximal independent set, for every $x \in X \sim Y$, there exists a $y \in Y$ such that the two elements x and y are not independent; this means either $x \in F(y)$ or $x \in F^{-1}(y)$ (or equivalently, $x \in F(y) \cup F^{-1}(y) = F^{\pm}(y)$).

A theorem concerning the available size of independent sets (under the set-valued mapping $F : X \dashrightarrow X$ with the self-independence property that for each $x \in X$, $x \notin F(x)$) has been proven by Erdős [1950]:

4.8 Theorem   *Let X be a set of infinite cardinality $|X| = m$. If there is a smaller cardinal number $n < m$ such that, for each $x \in X$, $|F(x)| < n$, then there exists an independent set $Y \subset X$ with cardinality $|Y| = m$.*

Erdős's theorem is valid for all infinite cardinals m, regular and singular. But for our purposes we only need the cases $m = \aleph_0$ (i.e. countably infinite sets, e.g. when $X = \mathbb{N}$) and $m = c$ (i.e. uncountably infinite sets with cardinality of the continuum, e.g. when $X = \mathbb{R}$). Since all subsets of an independent set are independent, when there is an independent set with cardinality m, there are also independent sets with cardinality k for all $k \leq m$.

Consider the example of a finite set $X = \{1, 2, 3\}$ and $F : X \dashrightarrow X$ with $F(1) = \{2\}$, $F(2) = \{3\}$, and $F(3) = \{1\}$. Then one sees that $|X| = 3$ and each $|F(x)| < 2 < 3$, but X itself (which is the only 3-element subset of X) is not independent (indeed, no 2-element subset of X is independent either). This shows that the theorem does not apply when the set X is of *finite cardinality*.

Also, as noted in Erdős's paper, the condition $|F(x)| < n < m$, the existence of the '*uniform bound*' n for all $F(x)$, cannot be weakened to $|F(x)| < m$. Consider $X = \mathbb{N}$, and define

(13)
$$\begin{cases} F(1) = \varnothing \\ F(k) = \{1, 2, ..., k-1\} \quad \text{for } k > 1 \end{cases}.$$

Here $|X| = \aleph_0$ and each $|F(x)| < \aleph_0$, but no infinite subset of X is independent. Indeed, besides trivially $\varnothing$ and singleton sets, no other finite or infinite subset of X is independent under this F.

In the counterexample (13), while each $F(x)$ is a finite set, the same cannot be said for $F^{-1}(x)$, since

(14) $F^{-1}(k) = \{k+1, k+2, k+3, ...,\}$ for all $k \in \mathbb{N}$.

This motivates the idea that perhaps if one also restricts $F^{-1}(x)$ to be finite, then the availability of infinite independent sets may be entailed. Indeed, one has the following:

4.9 Theorem   *Let X be a set of infinite cardinality $|X| = m$. If, for each $x \in X$, both $|F(x)| < m$ and $|F^{-1}(x)| < m$, then there exists an independent set $Y \subset X$ with cardinality $|Y| = m$.*

Proof   One may note that $|F(x)| < m$ and $|F^{-1}(x)| < m$ imply $|F^{\pm}(x)| = |F(x) \cup F^{-1}(x)| < m$ (m being infinite). Assume the contrary that X does not contain an independent set of cardinality m. Let $Y \subset X$ be a maximal independent set, so by assumption $|Y| < m$. Let $y \in Y$; on account of Lemma 4.5, both $F(y) \subset X \sim Y$ and $F^{-1}(y) \subset X \sim Y$; thus $F^{\pm}(y) \subset X \sim Y$, whence

(15) $\displaystyle\bigcup_{y \in Y} F^{\pm}(y) \subset X \sim Y.$

As explained in Section 4.7, the fact that Y is a maximal independent set means that for every $x \in X \sim Y$, there exists a $y \in Y$ such that $x \in F^{\pm}(y)$; thus $X \sim Y \subset \displaystyle\bigcup_{y \in Y} F^{\pm}(y)$, and together with (15) one has $\displaystyle\bigcup_{y \in Y} F^{\pm}(y)$ $= X \sim Y$.

This means

(16) $X = Y \cup \displaystyle\bigcup_{y \in Y} F^{\pm}(y).$

Now $|X| = m$ while $|Y| < m$. The equality (16) therefore implies for at least one $y \in Y$, $\left|F^{\pm}(y)\right| = m$ (which in turn implies either $\left|F(y)\right| = m$ or $\left|F^{-1}(y)\right| = m$), contradicting the hypothesis. $\square$

Again, all I need for my purposes from Theorem 4.9 are the two cases $m = \aleph_0$ and $m = c$. When Theorem 4.9 specializes into $m = \aleph_0$, it gives the

4.10 Corollary   *If X is a countably infinite set and if, for each $x \in X$, both $F(x)$ and $F^{-1}(x)$ are finite sets, then X contains a countably infinite independent set.*

Here, one sees that when both $F(x)$ and $F^{-1}(x)$ are bounded, the boundedness does not have to be uniform.

Counting Independent Sets

4.11 Definition   An undirected graph is called k *-colourable* if to each vertex one of a given set of k colours can be attached in such a way that on each edge the two endpoints are assigned different colours.

Note that if a graph is k -colourable, then it is n -colourable for all $n \geq k$ (up to the number of vertices of the graph). If k is the smallest integer for which a graph is k -colourable (i.e. the graph is k -colourable but not $(k-1)$ -colourable), then the graph is called k *-chromatic*.

Given a relation F on X, consider the symmetric closure of F, the mapping $F^{\pm} = F \cup F^{-1} : X \multimap X$ defined in (4) above. Two elements $x, y \in X$ are not independent iff $y \in F^{\pm}(x) = F(x) \cup F^{-1}(x)$ (*cf.* (7)), which means that in the undirected graph $F^{\pm}$, there is an edge joining vertices x and y, and in a colouring scheme, the two vertices must therefore be assigned different colours. Stated otherwise, $x, y \in X$ may be assigned different colours iff they are independent. Thus the graph $F^{\pm}$ is k -colourable iff the set X may be partitioned into k independent sets under the set-valued mapping $F^{\pm} : X \multimap X$ (with all members in each independent set receiving the same one of the k colours).

The colouring of the vertices of the graph $F^{\pm}$ also illustrates why the condition of 'self-independence', (11) and (12), is imposed. If an element $x \in X$ is not independent of itself, then there is an edge (a 'self-loop') that joins vertex x

to itself, implying that ' x and x ' must be assigned different colours, and therefore, vertex x cannot be coloured at all.

Erdős continued his exploration of independent sets in the paper de Bruijn and Erdős [1951], beginning with a general theorem on graph colouring:

4.12 Theorem   *Let k be a positive integer, and let the graph G have the property that any finite subgraph is k-colourable. Then G itself is k-colourable.*

Note that the Axiom of Choice (*cf.* Sections 0.20 and 1.2) is required for the proof of this theorem, whence also for the subsequent theorems that use this result.

F continues to be a relation on a set X, $F : X \multimap X$ with the self-independence property that for each $x \in X$, $x \notin F(x)$. Note that X is the disjoint union of its singleton sets:

$$(17) \qquad\qquad X = \bigcup_{x \in X} \{x\}.$$

Since each $\{x\}$ is an independent set, one sees that X may be partitioned into $|X|$ independent sets. The number $|X|$ is, of course, the maximum. In the simple example of the relation on X defined by $F(x) = X \sim \{x\}$, one sees that no two different elements are independent (i.e. the only independent sets under F are the singleton sets and the empty set), whence the upper bound $|X|$ of the number of independent sets in the partition of X cannot be lowered for this F.

Since all subsets of an independent set are independent (*cf.* Section 4.6), an independent set with two or more elements may be further partitioned into independent sets. Thus, if X may be partitioned into k independent sets, then it may be partitioned into n independent sets for all $n \geq k$ (up to $|X|$).

If two independent sets $Y, Z \subset X$ are such that $Y \cap F(Z) = \varnothing$ and $Z \cap F(Y) = \varnothing$, then they may be merged into one independent set $Y \cup Z \subset X$. So a question may be posed: Under what conditions may X be partitioned into fewer than $|X|$ independent sets? de Bruijn and Erdős's paper continues with the following:

4.13 Theorem   *Let X be a finite or countably infinite set. If there exists a non-negative integer n such that, for each $x \in X$, $|F(x)| \leq n$, then X may be partitioned into $2n+1$ independent sets.*

Proof   First consider the case when X is finite, and proceed through induction on $|X|$. The initial case for $|X| = 1$ is trivial. Now let $k > 1$, assume the theorem to be true for $|X| = k - 1$, and consider next $|X| = k$.

Since $|F(x)| \le n$ and $|X| = k$, the undirected graph defined by the symmetric closure $F^{\pm}$ of F can have at most kn edges. So there exists a vertex v that is connected with fewer than $2n+1$ vertices. (If every vertex is connected with $2n+1$ or more vertices, then the total number of edges in the graph $F^{\pm}$ would be $\ge k(2n+1)/2 > kn$.) The induction hypothesis posits that $X \sim \{v\}$ (with cardinality $k-1$) may be partitioned into $2n+1$ independent sets. Since the vertex v has fewer than $2n+1$ edges attached, it is not connected with at least one of these $2n+1$ independent sets of the partition, say Y. Thus $v \in X$ is independent of all elements of Y, whence $Y \cup \{v\}$ is an independent set. The remaining $2n$ independent sets together with $Y \cup \{v\}$ then form a partition of the set X with $|X| = k$, completing the induction.

Since the partition of X into $2n+1$ independent sets can be interpreted as $(2n+1)$-colourability of (finite subgraphs of) the undirected graph $F^{\pm}$, the countably infinite case follows from Theorem 4.12.   □

The exercise is to decrease the number of independent sets in the partition from the maximum $|X|$, so the theorem is only of interest when $2n+1 \le |X|$, i.e. when $n < \frac{1}{2}|X|$. If X is a countably infinite set, then at least one of these $2n+1$ independent subsets must be countably infinite. Thus Theorem 4.13 (a theorem on the *number* of independent sets) implies Theorem 4.8 (a theorem on the *size* of independent sets) for the case when $|X| = \aleph_0$. Note that both Theorems 4.8 and 4.13 require a *uniform bound* n for all $|F(x)| \le n$.

4.14 Corollary   *If X is a countably infinite set and, for each $x \in X$, $F(x)$ is a finite set, then X is the union of (i.e. may be partitioned into) a countable number of independent sets.*

Proof   For each k , define X_k as the set of all $x \in X$ for which $|F(x)| = k$. Then X is the disjoint union of the sets X_0, X_1, X_2, ... Apply Theorem 4.13 to each X_k .   □

4.15 Examples   Let $n > 0$ be an integer. Consider $F : \mathbb{Z} \multimap \mathbb{Z}$ defined for $k \in \mathbb{Z}$ (the set of all integers) by

(18) $$F(k) = \{k+1, k+2, ..., k+n\},$$

whence $|F(k)| = n$ $(\leq n)$, satisfying the hypothesis of Theorem 4.13. Then one may verify that the $2n+1$ numbers 0, 1, 2, ..., $2n$ are pairwise dependent each on the other (i.e. not pairwise independent). Indeed, $\mathbb{Z}$ may be partitioned into the $2n+1$ independent sets

$$
(19) \quad
\begin{aligned}
[k]_{2n+1} &= \{x \in Z : x \equiv k \pmod{2n+1}\} \\
&= \{..., -2(2n+1)+k, -(2n+1)+k, k, (2n+1)+k, 2(2n+1)+k, ...\},
\end{aligned}
$$
$$
k = 0, 1, ..., 2n,
$$

but no fewer. (See *ML*: 1.15 for a review of the relation *congruence modulo m* on $\mathbb{Z}$, $\mathbb{Z}_m = \mathbb{Z}/\equiv \pmod{m}$.) This shows the lower bound $2n+1$ is the best possible under the circumstances.

For example, when $n = 3$, F as defined in (18) partitions $\mathbb{Z}$ into $2n+1 = 7$ independent sets

$$
(20) \quad
\begin{aligned}
[0]_7 &= \{x \in \mathbb{Z} : x \equiv 0 \pmod{7}\} = \{..., -14, -7, 0, 7, 14, ...\}, \\
[1]_7 &= \{x \in \mathbb{Z} : x \equiv 1 \pmod{7}\} = \{..., -13, -6, 1, 8, 15, ...\}, \\
&\;\;\vdots \\
[6]_7 &= \{x \in \mathbb{Z} : x \equiv 6 \pmod{7}\} = \{..., -8, -1, 6, 13, 20, ...\};
\end{aligned}
$$

and the *digraph* $F : \mathbb{Z}_7 \multimap \mathbb{Z}_7$ appears thus:

(21)

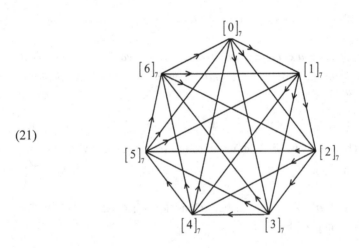

The associated undirected graph $F^{\pm} : \mathbb{Z}_7 \multimap \mathbb{Z}_7$ (i.e. diagram (21) without the arrows on the edges) is the complete graph K_7. (A *complete graph* is a graph in which any two distinct vertices are joined by an edge, i.e. one may travel from each vertex to every other vertex on a path of length one. Indeed, for a general $n > 0$, the undirected graph $F^{\pm} : \mathbb{Z}_{2n+1} \multimap \mathbb{Z}_{2n+1}$ is the complete graph K_{2n+1}, with $2n+1$ vertices and $n(2n+1)$ edges.)

It is important to note that while the lower bound $2n+1$ is attained for certain examples, Theorem 4.13 simply guarantees that under the condition of the uniform boundedness of $|F(x)| \leq n$, X may be partitioned into no more than $2n+1$ independent sets. It does not, in particular, forbid the partition of X into fewer than $2n+1$ independent sets. But of course, as noted earlier, if X may be partitioned into k independent sets and $k < 2n+1$, these independent sets may be further subdivided, so that X may be partitioned into m independent sets for all integer m within the range $k \leq m \leq |X|$; this includes the special case $m = 2n+1$.

As an illustration, consider $X = \mathbb{N}_0 = \{0\} \cup \mathbb{N} = \{0,1,2,3,...\}$ and $F : \mathbb{N}_0 \multimap \mathbb{N}_0$ defined by

(22)
$$\begin{cases} F(0) = \varnothing \\ F(k) = \{0, k+1\} \quad \text{for } k \geq 1 \end{cases}$$

One sees that, for each $k \geq 1$, the pair of elements 0 and k is not independent, since $0 \in F(k)$; thus $\{0\}$ is a maximal independent set. It is evident that the set $D = \{1,3,5,...\}$ of all odd natural numbers and the set $E = \{2,4,6,...\}$ of all even natural numbers are also maximal independent sets. So $\mathbb{N}_0$ may be partitioned into three independent sets, $\mathbb{N}_0 = \{0\} \cup D \cup E$. Since for each $x \in \mathbb{N}_0$, $|F(x)| \leq 2$, Theorem 4.13 applies and concludes that $\mathbb{N}_0$ may be partitioned into five independent sets. One sees that actually three would do (but of course, either D or E may be further partitioned, and $\mathbb{N}_0$ may indeed be partitioned into five independent sets; so this is not a counterexample to the proven theorem).

Note that Theorem 4.13 entails (and so does Theorem 4.8) that when X is countably infinite and $|F(x)|$ is uniformly bounded, one countably infinite set must exist, but it does not say that there must be a partition of X into $2n+1$ countably infinite independent sets. This is shown in our example (22), in which $\{0\}$ is a maximal independent set (whence it cannot be expanded to an infinite independent set). One also sees that in this example, no element $x \in \mathbb{N}_0$ is isolated, so this additional property does not guarantee the existence of a partition of X into all countably infinite independent sets either.

In example (22), while $|F(x)|$ is bounded, the set $F^{-1}(0) = \mathbb{N}$ is infinite. As we observed in the counterexample (13) and the subsequent Theorem 4.9 (and Corollary 4.10), perhaps here again if one also restricts $|F^{-1}(x)|$ to be bounded, then more may be entailed. Indeed, one may conclude further:

4.16 Theorem   *If X is a countably infinite set and if there exists a non-negative integer n such that, for each $x \in X$, $|F(x)| \leq n$ and $|F^{-1}(x)| \leq n$, then X may be partitioned as the finite union of countably infinite independent sets.*

  Proof   Theorem 4.13 partitions X into the finite union of independent sets, with at least one of these independent sets countably infinite. So let $Y \subset X$ be a countably infinite independent set. The remainder set $X \sim Y$ is itself a finite union of independent sets. Now suppose one of these independent sets, say Z, is finite. Choose $x_1 \in Z$; the bounds $|F(x_1)| \leq n$ and $|F^{-1}(x_1)| \leq n$ imply that $F^{\pm}(x_1) = F(x_1) \cup F^{-1}(x_1)$ is a finite set (with cardinality at most $2n$). This means the set $Y \cap F^{\pm}(x_1)$ is finite or, what is the same, the set $Y \sim F^{\pm}(x_1)$ is countably infinite; let it be enumerated thus:

(23) $$Y \sim F^{\pm}(x_1) = \{y_1, y_2, y_3, y_4, ...\}.$$

Now consider the two sets

(24) $$Y_1 = \{x_1, y_1, y_3, ...\}, \ Y_2 = Y \sim \{y_1, y_3, ...\},$$

i.e. every other element of $Y \sim F^{\pm}(x_1)$ is appended to $\{x_1\}$ to form the set Y_1, and $Y_2 = Y \sim Y_1$. In other words, the original two disjoint independent sets, the countably infinite Y and the finite Z, are repartitioned into the two countably infinite Y_1 (which contains x_1) and Y_2 and the finite $Z \sim \{x_1\}$, and each of the three sets is independent.

  The procedure may be iterated, by next considering $x_2 \in Z \sim \{x_1\}$ and appending it to 'half' of $Y_2 \sim F^{\pm}(x_2)$. Each iteration increases the number of countably infinite independent sets by one and decreases the number of remaining elements in Z by one. Since Z is finite, the procedure halts. If there are more finite independent sets in the partition of $X \sim Y$, the same procedure may be carried out. In sum, this splitting of a countably infinite set into two such, with the appending of one element to one 'half' of the split each time, may only be continued a finite number of times. At the end, X is partitioned into a finite number of countably infinite independent sets.   □

4.17 Another Example   The hypothesis of Theorem 4.16, that for each $x \in X$, $|F(x)| \le n < \aleph_0$ and $|F^{-1}(x)| \le n < \aleph_0$ (i.e. that there is a uniform finite bound for all sets concerned), is stronger than that of Corollary 4.10, which only requires that $|F(x)| < \aleph_0$ and $|F^{-1}(x)| < \aleph_0$ (i.e. that the sets only have to be bounded, but not necessarily uniformly bounded). The following is an example that shows the stronger condition cannot be weakened to the latter: The finiteness of all $F(x)$ and $F^{-1}(x)$ is not enough to guarantee the possible partitioning of X as the finite union of countably infinite independent sets.

Firstly, as noted in Section 4.12, under the relation on X defined by $F(x) = X \sim \{x\}$, no two different elements of X are independent, the only independent sets under this F being the singleton sets and the empty set. If X is a finite set, then it cannot be partitioned into fewer than $|X|$ independent sets (each of which is a singleton set). One may also note that for this F, $F^{-1}(x) = X \sim \{x\}$.

Next, I shall proceed to define a set-valued mapping $F: \mathbb{N} \multimap \mathbb{N}$. The *triangular numbers* $T_n = 1 + 2 + \cdots + n = \binom{n+1}{2}$, $n = 1, 2, 3, \ldots$, count the 'dots' that fill an equilateral triangle with n dots on a side:

$$(25) \qquad\qquad T_n = 1, 3, 6, 10, \ldots .$$

Now consider subsets of natural numbers from one triangular number to the next:

$$(26) \qquad \begin{cases} X_1 = \{T_1\} \\ X_n = \{T_{n-1} + 1, T_{n-1} + 2, , \ldots, T_n\} \quad \text{for } n \ge 2 \end{cases},$$

i.e.

$$(27) \qquad \begin{cases} X_1 = \{1\} \\ X_2 = \{2, 3\} \\ X_3 = \{4, 5, 6\} \\ \vdots \end{cases}.$$

So $\mathbb{N} = X_1 \cup X_2 \cup X_3 \cup \cdots$ and each $|X_n| = n$. Define

(28) $$F(x) = X_n \sim \{x\} \text{ if } x \in X_n \subset \mathbb{N}.$$

Evidently, all $F(x)$ and $F^{-1}(x)$ are finite sets. Since for each $n = 1, 2, 3, \ldots$, no two different elements of X_n are independent under F, an independent set $Y \subset \mathbb{N}$ can contain at most one element from each X_n. This means after partitioning off k countably infinite independent sets $Y_1, Y_2, \ldots, Y_k$ from $\mathbb{N}$, at most k elements from each X_n are contained in $\bigcup_{j=1}^{k} Y_j$; thus all the infinitely many 'remainder' sets

(29) $$X_n \sim \bigcup_{j=1}^{k} Y_j \text{ for } n > k$$

are still nonempty. This shows $\mathbb{N}$ cannot be partitioned into a finite number of independent sets.

4.18 Counting Maximal Independent Sets   Theorems 4.13 and 4.16 are examples where under certain finite-bound conditions on F, the domain X may be partitioned as the union of a finite number M of independent sets. This number M, then, provides an upper bound, thence an estimate, of the number of *maximal* independent sets under the set-valued mapping $F : X \multimap X$.

5
Set-Valued Mappings Redux

I will write on the tablets the words that were on the former
tablets, which you smashed, and you shall put them in the ark.

— *Deuteronomium* 10:2

This chapter contains a repetition and a commentary on many of the results in
the theory of set-valued mappings, recast in the formalism of matrix algebra, for
the purpose of further explication, illustration, and clarification. The sequence of
set-valued concepts in their matrix representations will appear below in
approximately the same order as that of their appearances in the previous four
chapters.

Si Postquam: Binary Matrices

5.1 Cartesian Product   Let X and Y be two finite sets with $|X| = m$ and
$|Y| = n$. Let

(1) $$\left\{x_i\right\}_{i \in \{1,2,...,m\}} = \left\{x_1, x_2, ..., x_m\right\} \quad \text{and} \quad \left\{y_j\right\}_{j \in \{1,2,...,n\}} = \left\{y_1, y_2, ..., y_n\right\}$$

be their respective (arbitrarily chosen but fixed) enumeration as sequences (*cf.*
Section 0.25).   Their Cartesian product $X \times Y$ may be represented by (i.e., is
isomorphic to) an $m \times n$ matrix M, with rows i ranging from 1 to m indexing
X and columns j ranging from 1 to n indexing Y.   The (i, j) th entry M_{ij} in
the matrix represents the ordered pair (x_i, y_j).

An analogous representation of $X \times Y$ as an 'infinite matrix' is also possible
when either $m = \aleph_0$ or $n = \aleph_0$ (or both).   That is, when one or both of the sets X
and Y are countably infinite (whence enumerated as infinite sequences), their
Cartesian product $X \times Y$ may be represented by an infinite matrix, if one makes

A.H. Louie, *The Reflection of Life*, IFSR International Series on Systems Science
and Engineering 29, DOI 10.1007/978-1-4614-6928-5_5,
© Springer Science+Business Media New York 2013

strategic uses of ellipses to indicate trends. Most of the finite-dimensional matrices appearing henceforth (and the statement concerning them) may naturally be extended to their countably infinite-dimensional counterparts.

In what follows, X and Y will, unless otherwise stated, be two finite sets with $|X| = m$ and $|Y| = n$, and indexed as in (1).

5.2 Adjacency Matrix   If $R \subset X \times Y$ is a relation between the finite sets X and Y, then R may be represented by the *adjacency matrix*, which I shall denote by $[R]$, with entries defined by

(2)
$$R_{ij} = \begin{cases} 1 & \text{if } (x_i, y_j) \in R \\ 0 & \text{if } (x_i, y_j) \notin R \end{cases}.$$

An adjacency matrix is an example of a binary matrix; it is also called a $(0,1)$-matrix since all its entries are either '0' or '1'.

The adjacency matrix $[f]$ of a mapping $f : X \to Y$ is a binary matrix with exactly one '1' in each row. Recall the notation of f as a mapping of X into Y (*cf.* Sections 1.6 and 1.24) implicitly defines that $\mathrm{dom}(f) = X$. The range of f is the collection of elements of Y for which the corresponding column in the adjacency matrix has at least one '1'. For a constant mapping, its adjacency matrix has its '1' on each row appearing in exactly the same column.

Adjacency Matrices of Mappings

5.3 Sur-/In-/Bijective Mappings   The adjacency matrix of a *surjective* mapping has at least one '1' in each column. The adjacency matrix of an *injective* mapping has at most one '1' in each column. The adjacency matrix of a *bijective* mapping, therefore, has exactly one '1' in each column.

The requirement that, in the adjacency matrix $[f]$ of a mapping $f : X \to Y$, there is exactly one '1' in each of the m rows of the $m \times n$ matrix, in fact, means that the adjacency matrix itself must contain exactly m '1's (with all the other entries filled by '0's). The 'density' of '1's in the matrix is thus $\dfrac{m}{mn} = \dfrac{1}{n}$, so $[f]$ is a *sparse matrix* for large n (noting that the quality of 'sparsity' is, however, relative). The 'exactly m '1's in $[f]$' implies that if $m < n$, then $f : X \to Y$ cannot be surjective; if $m > n$, then $f : X \to Y$ cannot be injective; and $m = n$ is a prerequisite for $f : X \to Y$ to be bijective.

5.4 Inverse Mapping When $f : X \to Y$ is injective, an inverse mapping $f^{-1} : \mathrm{ran}(f) \to X$ exists (*cf.* Section 1.15). Injectivity implies that $m \le n$, and $|\mathrm{ran}(f)| = |X| = m$.

The $m \times m$ adjacency matrix $\left[f^{-1} \right]$ of $f^{-1} : \mathrm{ran}(f) \to X$ may be constructed from the $m \times n$ adjacency matrix $[f]$ of $f : X \to Y$ as follows. Firstly, form the $n \times m$ *transpose* matrix $[f]^t$ of $[f]$ with entries defined by

(3) $f^t_{ij} = f_{ji}$ for $i = 1, 2, ..., n$ and $j = 1, 2, ..., m$.

That f is injective means $[f]$ has at most one '1' in each column, whence $[f]^t$ has at most one '1' in each row. So $[f]^t$ has not violated the 'exactly one '1' in each row' requirement of a mapping.

Next, remove all the $n - m$ rows of $[f]^t$ that consist entirely of '0's. These rows correspond to the '0' columns of $[f]$, that is, elements of $Y \sim \mathrm{ran}(f)$. The matrix $[f]^t$ resulting from $[f]^t$ after the removal of the '0' rows has dimension $m \times m$. $[f]^t$ is the adjacency matrix $\left[f^{-1} \right]$ of f^{-1}. Since $f : X \to Y$ is a mapping, matrix $[f]$ has exactly one '1' on each row, thence both matrices $[f]^t$ and $[f]^t = \left[f^{-1} \right]$ have exactly one '1' in each column. This simply illustrates the fact that f^{-1} is necessarily a *one-to-one* mapping of $f(X) = \mathrm{ran}(f)$ *onto* $X = \mathrm{dom}(f)$; that is, that $f^{-1} : \mathrm{ran}(f) \to X$ is a *bijection*.

5.5 Sequential Composition Let X, Y, and Z be finite sets with $|X| = m$, $|Y| = n$, and $|Z| = l$. Consider three mappings: $f : X \to Y$, $g : Z \to X$, and their sequential composite $f \circ g : Z \to Y$. Their adjacency matrices $[f]$, $[g]$, and $[f \circ g]$ have dimensions $m \times n$, $l \times m$, and $l \times n$, respectively.

The three matrices are related by a simple matrix multiplication:

(4) $[f \circ g] = [g][f]$

(note the reverse order, i.e., *contravariance*). The (k, j) th entry (where $1 \le k \le l$ and $1 \le j \le n$) in the matrix $[f \circ g]$ is

(5)
$$\left(f \circ g\right)_{kj} = \sum_{i=1}^{m} g_{ki} \, f_{ij} \, .$$

Since g is a mapping, the k th row of its adjacency matrix $[g]$, that is,

(6)
$$\begin{bmatrix} g_{k1} & g_{k2} & \cdots & g_{km} \end{bmatrix},$$

has exactly one '1' in it; all the other entries are '0'. So the sum appearing in (5) has only one possible nonzero term, namely, $g_{ki} \, f_{ij}$ for the index i with $g_{ki} = 1$, and $g_{ki} \, f_{ij} = 1$ only happens when $f_{ij} = 1$ as well. The equivalence

(7)
$$\left(f \circ g\right)_{kj} = g_{ki} \, f_{ij} = 1 \quad \text{iff} \quad g_{ki} = 1 \text{ and } f_{ij} = 1$$

may be explicated thus. $g_{ki} = 1$ is the statement $g\left(z_k\right) = x_i$, and $f_{ij} = 1$ is the statement $f\left(x_i\right) = y_j$. This is the only combination that would result in $f\left(g\left(z_k\right)\right) = y_j$, that is, $\left(f \circ g\right)_{kj} = 1$. (If $f\left(x_i\right) = y_{j'}$ for a different index $j' \neq j$, then it would mean $\left(f \circ g\right)_{kj'} = 1$ but $\left(f \circ g\right)_{kj} = 0$.)

The fact that the sum appearing in (5) has only one possible nonzero term also means that even when the 'common middle' set X is countably infinite (i.e., $|X| = m = \aleph_0$), there is no convergence issue involved, although (5) is then formally an 'infinite sum'.

Adjacency Matrices of Set-Valued Mappings

5.6 Set-Valued Mapping   A set-valued mapping $F : X \multimap Y$ is a relation $F \subset X \times Y$; thus it may be represented by its adjacency matrix. In each row of the adjacency matrix of a set-valued mapping, the number of '1's may be any number inclusively between 0 and $|Y| = n$. (So in this way it differs from the adjacency matrix of a standard mapping, which has exactly one '1' in each row.)

The corange of F is the collection of elements of X for which the corresponding row in the adjacency matrix has at least one '1'. The range of F is the collection of elements of Y for which the corresponding column in the adjacency matrix has at least one '1'.

The adjacency matrix of a constant set-valued mapping has all its rows identical.

5.7 Various Special Mappings (*cf.* Definition 2.7)   The adjacency matrix of a *surjective* set-valued mapping has at least one '1' in each column.   In the adjacency matrix of a *semi-single-valued* set-valued mapping, if two different rows have one '1' in a common column, then the two rows must have all their '1's in the same columns, that is, the two rows must have identical entries. In the adjacency matrix of an *injective* set-valued mapping, no two rows can have identical entries except rows that are entirely filled with '0's. Since the number of '1's is not limited to one per row, none of these special set-valued mappings puts a constraint on the relative sizes of m and n.

5.8 Illustration A numerical example is in order. Let $|X| = 4$ and $|Y| = 5$ so $X \times Y$ is a 4×5 matrix. The two enumerating sequences are formally

(8) $$X = \{x_1, x_2, x_3, x_4\} \quad \text{and} \quad Y = \{y_1, y_2, y_3, y_4, y_5\},$$

but any listing will do. For example, X may simply be the set $\{a, b, c, d\}$ indexed alphabetically, that is,

(9) $$\begin{cases} x_1 = x(1) = a \\ x_2 = x(2) = b \\ x_3 = x(3) = c \\ x_4 = x(4) = d \end{cases}.$$

Likewise, Y may simply be the set $\{1, 2, 3, 4, 5\}$ indexed numerically. Now consider a set-valued mapping $F : X \multimap Y$ defined as

(10) $$\begin{cases} F(a) = \{2, 3\} \\ F(b) = \varnothing \\ F(c) = \{2, 3\} \\ F(d) = \{1, 5\} \end{cases}.$$

Its adjacency matrix $[F]$ is

(11)
$$\begin{array}{c} \\ a \\ b \\ c \\ d \end{array} \begin{array}{ccccc} 1 & 2 & 3 & 4 & 5 \\ \left[\begin{array}{ccccc} 0 & 1 & 1 & 0 & 0 \\ 0 & 0 & 0 & 0 & 0 \\ 0 & 1 & 1 & 0 & 0 \\ 1 & 0 & 0 & 0 & 1 \end{array} \right] \end{array}.$$

(One does not usually need to provide the row and column 'headings' in the tabulation of the matrix array, just the $(0,1)$-matrix itself suffices. I am simply including them in this first example for illustrative clarity.)

The corange of this F is $\{a,c,d\}$ and, indeed, $F(b) = \varnothing$ is reflected in the entire row of '0's in row b. The range of this F is $\{1,2,3,5\}$, and column 4 consists of all '0's; this also shows that F is not surjective. Rows a and c are identical and the entries are not all '0's, so F is not injective. There are no rows with 'partial overlaps' of '1's; thus F is a semi-single-valued set-valued mapping. F defines a partition of its range $\operatorname{ran}(F)$ into the blocks $\{2,3\}$ and $\{1,5\}$; its domain X is correspondingly partitioned into the blocks $\{a,c\}$ and $\{d\}$ plus $F^{+1}(\varnothing) = \{b\}$ (cf. Section 2.7).

Matrix Operations

5.9 Inverse Set-Valued Mapping   Given a set-valued mapping $F : X \multimap Y$, its inverse set-valued mapping $F^{-1} : Y \multimap X$ always exists without further requirements on F. The adjacency matrix $\left[F^{-1}\right]$ of F^{-1} is simply the transpose of the adjacency matrix of F, with no further row elimination needed. So if the adjacency matrix of F has dimension $m \times n$, then the adjacency matrix of F^{-1} has dimension $n \times m$.

For example, the adjacency matrix of F^{-1}, when F is as defined in (10), is the transpose of (11):

$$(12) \qquad \left[F^{-1}\right] = \begin{bmatrix} 0 & 0 & 0 & 1 \\ 1 & 0 & 1 & 0 \\ 1 & 0 & 1 & 0 \\ 0 & 0 & 0 & 0 \\ 0 & 0 & 0 & 1 \end{bmatrix}.$$

Note that rows 2 and 3 of $\left[F^{-1}\right]$ are identical and the entries are not all '0's, and there are no rows with 'partial overlaps' of '1's; thus F^{-1} is also semi-single-valued (which illustrates Theorem 2.15.iii).

5.10 The Poset of Relations $\langle \mathcal{P}(X \times Y), \subset \rangle$ is a partially ordered set. Let $F : X \multimap Y$ and $G : X \multimap Y$ be two set-valued mappings, with their $m \times n$ adjacency matrices $[F]$ and $[G]$. One may define a partial order $\leq$ on adjacency matrices (of the same dimensions) functorially as

(13) $$[F] \leq [G] \quad \text{iff} \quad F \subset G.$$

It is evident that the partial order on adjacency matrices is equivalent to a component-wise order: $[F] \leq [G]$ iff for all $i = 1, 2, ..., m$, $j = 1, 2, ..., n$,

(14) $$F_{ij} \leq G_{ij}.$$

5.11 Union and Intersection Let $F : X \multimap Y$ and $G : X \multimap Y$ be two set-valued mappings, with their $m \times n$ adjacency matrices $[F]$ and $[G]$. Then the adjacency matrix $[F \cup G]$ of their union mapping (Definition 2.28.i) $F \cup G : X \multimap Y$ is related to its 'constituents' by

(15) $$[F \cup G] = [F] \vee [G],$$

where the binary OR operation $\vee$ is defined on matrices entry-wise by

(16) $$(F \cup G)_{ij} = F_{ij} \vee G_{ij}, \quad i = 1, 2, ..., m, \quad j = 1, 2, ..., n,$$

where the standard logical *disjunction* $\vee$ on the $(0,1)$ entries is

(17) $$u \vee v = \begin{cases} 0 & \text{if } u = 0 \text{ and } v = 0 \\ 1 & \text{otherwise} \end{cases}.$$

Similarly, the adjacency matrix $[F \cap G]$ of their intersection mapping (Definition 2.28.ii) $F \cap G : X \multimap Y$ is related to its 'constituents' by

(18) $$[F \cap G] = [F] \wedge [G],$$

where the binary AND operation $\wedge$ is defined on matrices entry-wise by

(19) $$(F \cap G)_{ij} = F_{ij} \wedge G_{ij}, \quad i = 1, 2, ..., m, \quad j = 1, 2, ..., n,$$

where the standard logical *conjunction* $\wedge$ on the $(0,1)$ entries is

(20) $$u \wedge v = \begin{cases} 1 & \text{if } u = 1 \text{ and } v = 1 \\ 0 & \text{otherwise} \end{cases}.$$

5.12 Sequential Composition   Let X, Y, and Z be finite sets with $|X| = m$, $|Y| = n$, and $|Z| = l$. Consider three set-valued mappings, $F : X \multimap Y$, $G : Z \multimap X$, and their sequential composite $F \circ G : Z \multimap Y$ (Definition 3.4.i). As in the standard mapping example (*cf.* Section 5.5), their adjacency matrices $[F]$, $[G]$, and $[F \circ G]$ have dimensions $m \times n$, $l \times m$, and $l \times n$, respectively.

The three matrices are related contravariantly by a binary operation $\circ$ in

(21) $$[F \circ G] = [G] \circ [F]$$

(I am using the same symbol $\circ$, and the context will make it clear whether it is the sequential composition of set-valued mappings or the binary operation between their adjacency matrices.), where the (k, j) th entry (where $1 \le k \le l$ and $1 \le j \le n$) in the matrix $[F \circ G]$ is

(22) $$\left(F \circ G \right)_{kj} = \bigvee_{i=1}^{m} \left(G_{ki} \wedge F_{ij} \right),$$

with the disjunction and conjunction operations $\vee$ and $\wedge$ as defined in (17) and (20).

5.13 Square Product   With the same sets X, Y, and Z and set-valued mappings $F : X \multimap Y$ and $G : Z \multimap X$, the square product $F \square G : Z \multimap Y$ (Definition 3.4.ii) also has an $l \times n$ adjacency matrix, and $[F \square G]$ is defined by

(23) $$[F \square G] = [G] \square [F],$$

where the (k, j) th entry (where $1 \le k \le l$ and $1 \le j \le n$) in the matrix $[F \square G]$ is

(24) $$\left(F \square G \right)_{kj} = \bigwedge_{i=1}^{m} \left(G_{ki} \rightarrow F_{ij} \right),$$

with the binary IF/THEN operation $\rightarrow$ on the $(0,1)$ entries defined by the standard logical *conditional* $\rightarrow$:

(25) $$u \rightarrow v = \begin{cases} 0 & \text{if } u = 1 \text{ and } v = 0 \\ 1 & \text{otherwise} \end{cases}.$$

It is an easy exercise to verify that when both the set-valued mappings $F : X \multimap Y$ and $G : Z \multimap X$ are single-valued (whence each row of their adjacency matrices has exactly one '1' in it), both compositions (22) and (24) reduce to (5).

5.14 Illustration Let $F: X \multimap Y$ be as in Section 5.8. Let $Z = \{p, q, r, s\}$ and $G: Z \multimap X$ be defined by

(26)
$$\begin{cases} G(p) = \{c\} \\ G(q) = \{a, b\} \\ G(r) = \{d\} \\ G(s) = \varnothing \end{cases}.$$

Its adjacency matrix is

(27)
$$[G] = \begin{bmatrix} 0 & 0 & 1 & 0 \\ 1 & 1 & 0 & 0 \\ 0 & 0 & 0 & 1 \\ 0 & 0 & 0 & 0 \end{bmatrix}.$$

The sequential composite $F \circ G : Z \multimap Y$ is

(28)
$$\begin{cases} F \circ G(p) = \bigcup\limits_{x \in G(p)} F(x) = F(c) = \{2, 3\} \\ F \circ G(q) = \bigcup\limits_{x \in G(q)} F(x) = F(a) \cup F(b) = \{2, 3\} \\ F \circ G(r) = \bigcup\limits_{x \in G(r)} F(x) = F(d) = \{1, 5\} \\ F \circ G(s) = \bigcup\limits_{x \in G(s)} F(x) = \bigcup\limits_{x \in \varnothing} F(x) = \varnothing \end{cases}$$

(note the $\bigcup\limits_{\varnothing} = \inf \mathcal{P} Y = \varnothing$ for the value $F \circ G(s)$, as explained in Section 3.4).

The values in (28) are verified by the matrix binary operation formula $[F \circ G] = [G] \circ [F]$ via (22) in

(29)
$$[F \circ G] = \begin{bmatrix} 0 & 0 & 1 & 0 \\ 1 & 1 & 0 & 0 \\ 0 & 0 & 0 & 1 \\ 0 & 0 & 0 & 0 \end{bmatrix} \circ \begin{bmatrix} 0 & 1 & 1 & 0 & 0 \\ 0 & 0 & 0 & 0 & 0 \\ 0 & 1 & 1 & 0 & 0 \\ 1 & 0 & 0 & 0 & 1 \end{bmatrix} = \begin{bmatrix} 0 & 1 & 1 & 0 & 0 \\ 0 & 1 & 1 & 0 & 0 \\ 1 & 0 & 0 & 0 & 1 \\ 0 & 0 & 0 & 0 & 0 \end{bmatrix}.$$

Perhaps an illustration of how (29) is computed would be helpful; so consider the $(2,3)$ th entry of $[F \circ G]$:

$$(F \circ G)_{23} = \bigvee_{i=1}^{m} (G_{2i} \wedge F_{i3}) = \begin{bmatrix} 1 & 1 & 0 & 0 \end{bmatrix} \wedge \begin{bmatrix} 1 \\ 0 \\ 1 \\ 0 \end{bmatrix}$$

(30)
$$= (1 \wedge 1) \vee (1 \wedge 0) \vee (0 \wedge 1) \vee (0 \wedge 0)$$
$$= 1 \vee 0 \vee 0 \vee 0 = 1$$

Note that $(F \circ G)_{23} = 1$ means $y_3 \in F \circ G(z_2)$, that is, $3 \in F \circ G(q)$.

Similarly, one may compute the square product $F \square G : Z \dashv\!\mathrm{C}\, Y$ as

(31)
$$\begin{cases} F \square G(p) = \bigcap_{x \in G(p)} F(x) = F(c) = \{2,3\} \\[2mm] F \square G(q) = \bigcap_{x \in G(q)} F(x) = F(a) \cap F(b) = \{2,3\} \cap \varnothing = \varnothing \\[2mm] F \square G(r) = \bigcap_{x \in G(r)} F(x) = F(d) = \{1,5\} \\[2mm] F \square G(s) = \bigcap_{x \in G(s)} F(x) = \bigcap_{x \in \varnothing} F(x) = Y = \{1,2,3,4,5\} \end{cases}$$

(note the $\bigcap\limits_{\varnothing} = \sup PY = Y$ for the value $F \square G(s)$, as explained in Section 3.4).

The values in (31) are verified by the matrix binary operation formula $[F \square G] = [G] \square [F]$ via (24) in

(32)
$$[F \square G] = \begin{bmatrix} 0 & 0 & 1 & 0 \\ 1 & 1 & 0 & 0 \\ 0 & 0 & 0 & 1 \\ 0 & 0 & 0 & 0 \end{bmatrix} \square \begin{bmatrix} 0 & 1 & 1 & 0 & 0 \\ 0 & 0 & 0 & 0 & 0 \\ 0 & 1 & 1 & 0 & 0 \\ 1 & 0 & 0 & 0 & 1 \end{bmatrix} = \begin{bmatrix} 0 & 1 & 1 & 0 & 0 \\ 0 & 0 & 0 & 0 & 0 \\ 1 & 0 & 0 & 0 & 1 \\ 1 & 1 & 1 & 1 & 1 \end{bmatrix}.$$

Likewise, as an illustration of how (32) is computed, consider the $(2,3)$ th entry of $[F \square G]$:

$$(F \circ G)_{23} = \bigwedge_{i=1}^{m} (G_{2i} \to F_{i3}) = \begin{bmatrix} 1 & 1 & 0 & 0 \end{bmatrix} \to \begin{bmatrix} 1 \\ 0 \\ 1 \\ 0 \end{bmatrix}$$

(33)
$$= (1 \to 1) \wedge (1 \to 0) \wedge (0 \to 1) \wedge (0 \to 0) \quad ;$$
$$= 1 \wedge 0 \wedge 1 \wedge 1 = 0$$

note that $(F \circ G)_{23} = 0$ means $y_3 \notin F \circ G(z_2)$, that is, $3 \notin F \circ G(q)$. From this illustration, one may see that if row k of $[G]$ contains more than one '1's, say at columns $i_1, i_2, \dots$, then in column j of $[F]$, all the entries $(i_1, j), (i_2, j), \dots$ must also be '1's to ensure that $1 \to 0$ does not occur, whence entailing $(F \circ G)_{kj} = 1$.

This is the essence of the conjunction $\bigwedge_{i=1}^{m} w_i$, for which one component $w_i = 0$ will make the whole conjunction '0'; this is the logical equivalence of the intersection $\bigcap_{x \in G(z_k)} F(x)$, for which $y \in \bigcap_{x \in G(z_k)} F(x)$ iff $y \in F(x)$ for every $x \in G(z_k)$.

From, for example, $1 = (F \circ G)_{11} > (F \circ G)_{11} = 0$, one sees that $[F \circ G] \not\leq [F \circ G]$, whence illustrating that one does not necessarily have $F \circ G \subset F \circ G$ (cf. Section 3.4).

Relations on X

Let X remain a finite set with $|X| = m$ indexed as in (1). Let us now concentrate on relations on X, that is, set-valued mappings in the hom-set **Rel**(X, X) (Definition 3.13), with square adjacency matrices of dimension $m \times m$.

5.15 Various Special Mappings The adjacency matrix $[I]$ of the equality relation I on X (i.e., the identity morphism $1_X \in \mathbf{Rel}(X, X)$), $I : X \multimap X$, is, naturally, the $m \times m$ identity matrix, with a '1' on each of the m diagonal entries and '0' elsewhere. The matrix $[I]$ is often simply denoted by I, the same symbol for its mapping, without confusion.

Let $A \subset X$ and suppose $|A| = l \leq m$. Let the indexed elements of $A = \{a_1, a_2, \dots, a_l\}$ match the elements of $X = \{x_1, x_2, \dots, x_m\}$ by the correspondence

$a_j = x_{\sigma(j)}$ for $j = 1, 2, ..., l$ defined by an injective (single-valued) mapping $\sigma : \{1, 2, ..., l\} \rightarrow \{1, 2, ..., m\}$ (thus $\{a_j\}_{j \in \{1,2,...,l\}}$ is a subsequence of $\{x_i\}_{i \in \{1,2,...,m\}}$; *cf.* Section 0.24). Then the adjacency matrix $[i]$ of the inclusion map $i : A \subset\!\!\!-\, X$ of A in X is an $l \times m$ matrix that has a '1' at the l entries $(j, \sigma(j))$, $j = 1, 2, ..., l$, and '0' elsewhere. It is often convenient to permute the indexing, so that A consists of the first n elements of X (i.e., $\sigma(j) = j$ for $j = 1, 2, ..., l$), whence the adjacency matrix $[i]$ contains the $l \times l$ identity matrix $[1_A]$ as its initial block of l columns, and then fills out the remaining $m - l$ columns with '0's. Indeed, the adjacency matrix $[i]$ is often expanded to an $m \times m$ matrix, with the $l \times l$ identity matrix $[1_A]$ as its initial square block and then filled out to $m \times m$ with '0's. (Such an adjacency matrix $[i]$ corresponds to the set-valued mapping $i : X \subset\!\!\!-\, X$ defined by $i(x) = \{x\}$ for $x \in A$ and $i(x) = \varnothing$ for $x \in X \sim A$.) The same symbol $[1_A]$ may also denote the $m \times m$ matrix.

A set-valued mapping $R : X \subset\!\!\!-\, X$ is reflexive if and only if all the diagonal entries of its adjacency matrix $[R]$ are '1's. This means $I \leq [R]$ and clearly illustrates that R is reflexive iff $I \subset R$ (*cf.* Lemma 3.16(*r*)).

A set-valued mapping $R : X \subset\!\!\!-\, X$ is symmetric if and only if its adjacency matrix $[R]$ is symmetric, that is, $[R] = [R]^t = [R^{-1}]$ (*cf.* Lemma 3.16(*s*)).

A set-valued mapping $R : X \subset\!\!\!-\, X$ is antisymmetric if and only if, for $i, j = 1, 2, ..., m$,

(34) $R_{ij} = 1$ and $R_{ji} = 1 \Rightarrow i = j$,

whence iff $[R] \wedge [R^{-1}] \leq I$, iff $R \cap R^{-1} \subset I$ (*cf.* Lemma 3.16(*a*)).

A set-valued mapping $R : X \subset\!\!\!-\, X$ is transitive iff, for $i, j, k = 1, 2, ..., m$,

(35) $R_{ki} = 1$ and $R_{ij} = 1 \Rightarrow R_{kj} = 1$.

Now if $R_{ki} = 1$ and $R_{ij} = 1$, then $R_{ki} \wedge R_{ij} = 1$, whence from (22) one has

(36) $$(R \circ R)_{kj} = \bigvee_{i=1}^{m} (R_{ki} \wedge R_{ij}) = 1.$$

(In the disjunction $\bigvee\limits_{i=1}^{m} w_i$, one component $w_i = 1$ makes the whole disjunction '1'.) Thus, R is transitive iff $[R \circ R] \leq [R]$, iff $R \circ R \subset R$ (*cf.* Lemma 3.16(*t*)).

The adjacency matrix $[\varnothing]$ of the empty relation $\varnothing$ on X (i.e., the constant empty-set-valued mapping) is the $m \times m$ zero matrix, with a '0' in all of its m^2 entries. The adjacency matrix $[U]$ of the universal relation U on X is the $m \times m$ matrix with a '1' in all of its m^2 entries. It is evident that for all $R \in \mathbf{Rel}(X,X)$, $\varnothing \subset R \subset U$ whence $[\varnothing] \leq [R] \leq [U]$.

5.16 Isolation and Independence   An element $x_i \in X$ is isolated under the set-valued mapping $F : X \multimap X$ (Definition 4.2) iff in the adjacency matrix $[F]$, both the i th column and the i th row are filled entirely with '0's. Two elements $x_i, x_j \in X$ are independent under the set-valued mapping $F : X \multimap X$ (Definition 4.3) iff $F_{ij} = F_{ji} = 0$. The prescription of self-independence under the set-valued mapping $F : X \multimap X$ of all elements $x_i \in X$ (*cf.* Section 4.6) means that all the diagonal entries of such $[F]$ are '0's, that is, for each of $i = 1, 2, ..., m$, $F_{ii} = 0$.

Graphic Illustrations

5.17 The Simplest (M,R)-Network   Consider the digraph G

(37)

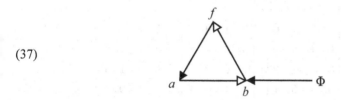

which the reader will recognize as that of the connected metabolism and repair components of the simplest (M,R)-network. (We may ignore the distinction of solid-headed and hollow-headed arrows for now.)   The set of vertices is $X = \{a, b, f, \Phi\}$ and the digraph G may be represented as the set-valued mapping $G : X \multimap X$ defined by

(38)
$$\begin{cases} G(a) = \{b\} \\ G(b) = \{f\} \\ G(f) = \{a\} \\ G(\Phi) = \{b\} \end{cases},$$

with $\text{dom}(G) = \text{cod}(G) = \text{cor}(G) = X = \{a,b,f,\Phi\}$, $\text{ran}(G) = \{a,b,f\}$, and adjacency matrix

(39)
$$[G] = \begin{bmatrix} 0 & 1 & 0 & 0 \\ 0 & 0 & 1 & 0 \\ 1 & 0 & 0 & 0 \\ 0 & 1 & 0 & 0 \end{bmatrix}.$$

Two arrows, initiating from the different vertices a and Φ , converge on the same terminating vertex b . This is reflected in the fact that $G(a) \cap G(\Phi)$ $= \{b\} \neq \varnothing$ and that the 'b'- column contains two '1's; thus G is not injective.

5.18 Reflexive Closure   The reflexive closure of G adds, for each $x \in X$, $x \in G(x)$, whence

(40)
$$\begin{cases} G^{=}(a) = \{a,b\} \\ G^{=}(b) = \{b,f\} \\ G^{=}(f) = \{a,f\} \\ G^{=}(\Phi) = \{b,\Phi\} \end{cases}.$$

It is $G^{=} = I \cup G$ (*cf.* Lemma 3.26), whence

(41)
$$\begin{aligned} \left[G^{=}\right] = \left[I \cup G\right] = I \vee [G] \\ = \begin{bmatrix} 1 & 0 & 0 & 0 \\ 0 & 1 & 0 & 0 \\ 0 & 0 & 1 & 0 \\ 0 & 0 & 0 & 1 \end{bmatrix} \vee \begin{bmatrix} 0 & 1 & 0 & 0 \\ 0 & 0 & 1 & 0 \\ 1 & 0 & 0 & 0 \\ 0 & 1 & 0 & 0 \end{bmatrix} = \begin{bmatrix} 1 & 1 & 0 & 0 \\ 0 & 1 & 1 & 0 \\ 1 & 0 & 1 & 0 \\ 0 & 1 & 0 & 1 \end{bmatrix}. \end{aligned}$$

5.19 'Inverse'   The inverse digraph $G^{-1} : X \multimap X$ of the two-component (M,R)-network (37) is defined by

(42)
$$\begin{cases} G^{-1}(a) = \{f\} \\ G^{-1}(b) = \{a,\Phi\} \\ G^{-1}(f) = \{b\} \\ G^{-1}(\Phi) = \varnothing \end{cases},$$

with $\operatorname{dom}(G^{-1}) = \operatorname{cod}(G^{-1}) = \operatorname{ran}(G^{-1}) = X = \{a,b,f,\Phi\}$, $\operatorname{cor}(G^{-1}) = \{a,b,f\}$, and adjacency matrix

$$(43) \qquad \left[G^{-1}\right] = [G]' = \begin{bmatrix} 0 & 0 & 1 & 0 \\ 1 & 0 & 0 & 1 \\ 0 & 1 & 0 & 0 \\ 0 & 0 & 0 & 0 \end{bmatrix}.$$

Note that there is at most one '1' in each column, indicating that G^{-1} is injective.

The two set-valued mappings G and G^{-1} compose thus:

$$(44) \quad \left[G^{-1} \circ G\right] = [G] \circ \left[G^{-1}\right] = \begin{bmatrix} 0 & 1 & 0 & 0 \\ 0 & 0 & 1 & 0 \\ 1 & 0 & 0 & 0 \\ 0 & 1 & 0 & 0 \end{bmatrix} \circ \begin{bmatrix} 0 & 0 & 1 & 0 \\ 1 & 0 & 0 & 1 \\ 0 & 1 & 0 & 0 \\ 0 & 0 & 0 & 0 \end{bmatrix} = \begin{bmatrix} 1 & 0 & 0 & 1 \\ 0 & 1 & 0 & 0 \\ 0 & 0 & 1 & 0 \\ 1 & 0 & 0 & 1 \end{bmatrix},$$

$$(45) \quad \left[G \circ G^{-1}\right] = \left[G^{-1}\right] \circ [G] = \begin{bmatrix} 0 & 0 & 1 & 0 \\ 1 & 0 & 0 & 1 \\ 0 & 1 & 0 & 0 \\ 0 & 0 & 0 & 0 \end{bmatrix} \circ \begin{bmatrix} 0 & 1 & 0 & 0 \\ 0 & 0 & 1 & 0 \\ 1 & 0 & 0 & 0 \\ 0 & 1 & 0 & 0 \end{bmatrix} = \begin{bmatrix} 1 & 0 & 0 & 0 \\ 0 & 1 & 0 & 0 \\ 0 & 0 & 1 & 0 \\ 0 & 0 & 0 & 0 \end{bmatrix}.$$

Lemma 2.21 states that for a set-valued mapping $F : X \multimap Y$, $A \subset F^{-1}(F(A))$ iff $A \subset \operatorname{cor}(F)$, which implies $1_{\operatorname{cor}(F)} \subset F^{-1} \circ F \subset X \times X$ (where the identity $1_{\operatorname{cor}(F)} : \operatorname{cor}(F) \multimap \operatorname{cor}(F)$ is also the inclusion map of $\operatorname{cor}(F)$ in X). In terms of adjacency matrices, this says

$$(46) \qquad \left[1_{\operatorname{cor}(F)}\right] \leq \left[F^{-1} \circ F\right]$$

(*cf.* Section 5.15 for the notation $[1_A]$). Both (44) and (45) verify this ordering (46): $\left[1_{\operatorname{cor}(G)}\right] \leq \left[G^{-1} \circ G\right]$ where $\left[1_{\operatorname{cor}(G)}\right]$ is the 4×4 identity matrix, since $\operatorname{cor}(G) = \{a,b,f,\Phi\}$, and $\left[1_{\operatorname{cor}(G^{-1})}\right] \leq \left[G \circ G^{-1}\right]$ where $\left[1_{\operatorname{cor}(G^{-1})}\right]$ is the 4×4 matrix with an initial block of a 3×3 identity matrix and filled out to 4×4 with '0's, since $\operatorname{cor}(G^{-1}) = \{a,b,f\}$.

Lemma 2.24 states that a set-valued mapping $F : X \rightarrowtail Y$ is injective iff $A = F^{-1}(F(A))$ for all $A \subset \mathrm{cor}(F)$, which means that the restriction $F^{-1} \circ F :$ $\mathrm{cor}(F) \rightarrowtail \mathrm{cor}(F)$ is the identity $1_{\mathrm{cor}(F)}$ and that the restriction $F^{-1} \circ F :$ $X \sim \mathrm{cor}(F) \rightarrowtail X$ is the empty mapping $\varnothing$; equivalently, $\left[F^{-1} \circ F \right] = \left[1_{\mathrm{cor}(F)} \right]$. The matrix $\left[G \circ G^{-1} \right]$ in (45) illustrates this: the initial 3×3 matrix is the identity (since the injective G^{-1} has $\mathrm{cor}(G^{-1}) = \{a, b, f\}$) and that the ' Φ '- row consists of all '0's; that is, $\left[G \circ G^{-1} \right] = \left[1_{\mathrm{cor}(G^{-1})} \right]$.

The matrices $\left[G^{-1} \circ G \right]$ and $\left[G \circ G^{-1} \right]$ in (44) and (45) also illustrate that $G \circ G^{-1} \neq G^{-1} \circ G$ and that neither composite is equal to $1_X = G^0$, so G^{-1} is not a true 'inverse' mapping of G. This illustrates that the 'power rule' $G^i \circ G^j = G^j \circ G^i = G^{i+j}$ does not cross the 'zero boundary', whence the two commutative power submonoids $\left\langle \{G^i : i \in \mathbb{N}_0\}, \circ \right\rangle$ and $\left\langle \{G^{-i} : i \in \mathbb{N}_0\}, \circ \right\rangle$ do not merge into an abelian group (*cf.* Section 3.22).

5.20 Symmetric Closure   The symmetric closure $G^{\pm}$ of G, the underlying undirected graph $G^{\pm} = G \cup G^{-1} : X \rightarrowtail X$ (*cf.* Lemma 3.27), is

(47)
$$\begin{cases} G^{\pm}(a) = \{b, f\} \\ G^{\pm}(b) = \{a, f, \Phi\} \\ G^{\pm}(f) = \{a, b\} \\ G^{\pm}(\Phi) = \{b\} \end{cases},$$

with the symmetric adjacency matrix

(48)
$$\left[G^{\pm} \right] = \left[G \cup G^{-1} \right] = \left[G \right] \vee \left[G^{-1} \right]$$
$$= \begin{bmatrix} 0 & 1 & 0 & 0 \\ 0 & 0 & 1 & 0 \\ 1 & 0 & 0 & 0 \\ 0 & 1 & 0 & 0 \end{bmatrix} \vee \begin{bmatrix} 0 & 0 & 1 & 0 \\ 1 & 0 & 0 & 1 \\ 0 & 1 & 0 & 0 \\ 0 & 0 & 0 & 0 \end{bmatrix} = \begin{bmatrix} 0 & 1 & 1 & 0 \\ 1 & 0 & 1 & 1 \\ 1 & 1 & 0 & 0 \\ 0 & 1 & 0 & 0 \end{bmatrix} .$$

5.21 Powers   Continuing with the same set-valued mapping $G : X \rightarrowtail X$, one may compute iteratively its powers (*cf.* Section 3.21). The square is

$$\left[G^2\right]=\left[G\circ G\right]=\left[G\right]\circ\left[G\right]$$

(49)
$$=\begin{bmatrix} 0 & 1 & 0 & 0 \\ 0 & 0 & 1 & 0 \\ 1 & 0 & 0 & 0 \\ 0 & 1 & 0 & 0 \end{bmatrix}\circ\begin{bmatrix} 0 & 1 & 0 & 0 \\ 0 & 0 & 1 & 0 \\ 1 & 0 & 0 & 0 \\ 0 & 1 & 0 & 0 \end{bmatrix}=\begin{bmatrix} 0 & 0 & 1 & 0 \\ 1 & 0 & 0 & 0 \\ 0 & 1 & 0 & 0 \\ 0 & 0 & 1 & 0 \end{bmatrix}.$$

Graphically, the adjacency matrix $\left[G^2\right]$ lists all paths of length 2 in the digraph G of diagram (37). For example, vertex f is reachable from vertex a after travelling on two connected edges, since $\left(G^2\right)_{13}=1$.

Next,

$$\left[G^3\right]=\left[G\circ G^2\right]=\left[G^2\right]\circ\left[G\right]$$

(50)
$$=\begin{bmatrix} 0 & 0 & 1 & 0 \\ 1 & 0 & 0 & 0 \\ 0 & 1 & 0 & 0 \\ 0 & 0 & 1 & 0 \end{bmatrix}\circ\begin{bmatrix} 0 & 1 & 0 & 0 \\ 0 & 0 & 1 & 0 \\ 1 & 0 & 0 & 0 \\ 0 & 1 & 0 & 0 \end{bmatrix}=\begin{bmatrix} 1 & 0 & 0 & 0 \\ 0 & 1 & 0 & 0 \\ 0 & 0 & 1 & 0 \\ 1 & 0 & 0 & 0 \end{bmatrix}.$$

Since the power submonoid $\left\langle\left\{G^i:i\in\mathbb{N}_0\right\},\circ\right\rangle$ is commutative (*cf.* Section 3.21), one may also calculate $\left[G^3\right]$ as $\left[G^3\right]=\left[G\right]\circ\left[G^2\right]$. In $\left[G^3\right]$, one sees that three diagonal entries are '1': $\left(G^3\right)_{11}=\left(G^3\right)_{22}=\left(G^3\right)_{33}=1$, that is, $x\in G^3(x)$ for $x=a,b,f$. So these three vertices are each reachable from itself after travelling on three edges whence they are each involved in a cycle (not necessarily the same cycle, but it happens to be the case in this example).

Continuing,

$$\left[G^4\right]=\left[G\circ G^3\right]=\left[G\right]\circ\left[G^3\right]$$

(51)
$$=\begin{bmatrix} 0 & 1 & 0 & 0 \\ 0 & 0 & 1 & 0 \\ 1 & 0 & 0 & 0 \\ 0 & 1 & 0 & 0 \end{bmatrix}\circ\begin{bmatrix} 1 & 0 & 0 & 0 \\ 0 & 1 & 0 & 0 \\ 0 & 0 & 1 & 0 \\ 1 & 0 & 0 & 0 \end{bmatrix}=\begin{bmatrix} 0 & 1 & 0 & 0 \\ 0 & 0 & 1 & 0 \\ 1 & 0 & 0 & 0 \\ 0 & 1 & 0 & 0 \end{bmatrix},$$

and one sees that in fact $\left[G^4\right]=\left[G\right]$! Vertices reachable after travelling on four edges are those that are reachable after travelling on one edge, that is, reachable directly. Note that although $G^4=G$, one cannot proceed to conclude that

$G^{-1} \circ G^4 = G^{-1} \circ G$ whence $G^3 = 1_X$. This is because G^{-1} is not a true 'inverse' mapping of G, as shown in Section 5.19.

5.22 Transitive Closure   One sees then that there are only three iterated positive powers of G; for $k = 1, 2, 3, \ldots$, G^k simply cycles through the three-member set $\{G, G^2, G^3\}$. This means the infinite union of the transitive closure (*cf*. Lemma 3.28) is the same as its truncation after three terms:

(52) $$\vec{G} = G \cup G^2 \cup G^3 \cup G^4 \cup \cdots = G \cup G^2 \cup G^3;$$

so

(53)
$$\left[\vec{G}\right] = \left[G \cup G^2 \cup G^3\right] = [G] \vee \left[G^2\right] \vee \left[G^3\right]$$
$$= \begin{bmatrix} 0 & 1 & 0 & 0 \\ 0 & 0 & 1 & 0 \\ 1 & 0 & 0 & 0 \\ 0 & 1 & 0 & 0 \end{bmatrix} \vee \begin{bmatrix} 0 & 0 & 1 & 0 \\ 1 & 0 & 0 & 0 \\ 0 & 1 & 0 & 0 \\ 0 & 0 & 1 & 0 \end{bmatrix} \vee \begin{bmatrix} 1 & 0 & 0 & 0 \\ 0 & 1 & 0 & 0 \\ 0 & 0 & 1 & 0 \\ 1 & 0 & 0 & 0 \end{bmatrix} = \begin{bmatrix} 1 & 1 & 1 & 0 \\ 1 & 1 & 1 & 0 \\ 1 & 1 & 1 & 0 \\ 1 & 1 & 1 & 0 \end{bmatrix}.$$

The first three columns of $\left[\vec{G}\right]$ are filled with '1's; this clearly illustrates that in the digraph G, vertices $\{a, b, f\}$ form a cycle, and this cycle is reachable from vertex Φ. The last column of $\left[\vec{G}\right]$ has all '0's, showing that the vertex Φ is not reachable from any vertex of G. This is, of course, the statement that in the two-component (M,R)-network (37), the 'repair' mapping Φ is not functionally entailed, a 'replication' mapping being what is required to complete the entailment. To make an (M,R)-network closed to efficient causation, one must 'repair the repair'.

5.23 Reachability   The reflexive transitive closure $\hat{G} = I \cup \vec{G}$ of G (*cf*. Section 3.36) is

(54)
$$\left[\hat{G}\right] = I \vee \left[\vec{G}\right] \left(= I \vee [G] \vee \left[G^2\right] \vee \left[G^3\right]\right)$$
$$= \begin{bmatrix} 1 & 0 & 0 & 0 \\ 0 & 1 & 0 & 0 \\ 0 & 0 & 1 & 0 \\ 0 & 0 & 0 & 1 \end{bmatrix} \vee \begin{bmatrix} 1 & 1 & 1 & 0 \\ 1 & 1 & 1 & 0 \\ 1 & 1 & 1 & 0 \\ 1 & 1 & 1 & 0 \end{bmatrix} = \begin{bmatrix} 1 & 1 & 1 & 0 \\ 1 & 1 & 1 & 0 \\ 1 & 1 & 1 & 0 \\ 1 & 1 & 1 & 1 \end{bmatrix}.$$

That the three vertices $\{a, b, f\}$ form a cycle in the digraph G, whence reachable from one another, has already been demonstrated by $\left[\vec{G}\right] \le \left[\hat{G}\right]$. The last column of $\left[\hat{G}\right]$ has all '0' entries except $\hat{G}_{44} = 1$; this shows the vertex Φ is not reachable from any vertex of G other than itself.

5.24 Connectedness   One may, likewise, compute iteratively the powers of the symmetric closure $G^{\pm}$ of G. Firstly,

$$\left[\left(G^{\pm}\right)^2\right] = \left[G^{\pm} \circ G^{\pm}\right] = \left[G^{\pm}\right] \circ \left[G^{\pm}\right]$$

(55)
$$= \begin{bmatrix} 0 & 1 & 1 & 0 \\ 1 & 0 & 1 & 1 \\ 1 & 1 & 0 & 0 \\ 0 & 1 & 0 & 0 \end{bmatrix} \circ \begin{bmatrix} 0 & 1 & 1 & 0 \\ 1 & 0 & 1 & 1 \\ 1 & 1 & 0 & 0 \\ 0 & 1 & 0 & 0 \end{bmatrix} = \begin{bmatrix} 1 & 1 & 1 & 1 \\ 1 & 1 & 1 & 0 \\ 1 & 1 & 1 & 1 \\ 1 & 0 & 1 & 1 \end{bmatrix}.$$

One sees that almost all entries of $\left[\left(G^{\pm}\right)^2\right]$ are '1', except $\left(G^{\pm}\right)^2_{24} = \left(G^{\pm}\right)^2_{42} = 0$.

This says in the underlying undirected graph $G^{\pm}$, all vertices are reachable from one another in a path of length 2, except between vertices b and Φ. Next,

$$\left[\left(G^{\pm}\right)^3\right] = \left[G^{\pm} \circ \left(G^{\pm}\right)^2\right] = \left[G^{\pm}\right] \circ \left[\left(G^{\pm}\right)^2\right]$$

(56)
$$= \begin{bmatrix} 0 & 1 & 1 & 0 \\ 1 & 0 & 1 & 1 \\ 1 & 1 & 0 & 0 \\ 0 & 1 & 0 & 0 \end{bmatrix} \circ \begin{bmatrix} 1 & 1 & 1 & 1 \\ 1 & 1 & 1 & 0 \\ 1 & 1 & 1 & 1 \\ 1 & 0 & 1 & 1 \end{bmatrix} = \begin{bmatrix} 1 & 1 & 1 & 1 \\ 1 & 1 & 1 & 1 \\ 1 & 1 & 1 & 1 \\ 1 & 1 & 1 & 0 \end{bmatrix}.$$

This says in $G^{\pm}$, all vertices are reachable from one another in a path of length 3, except from Φ to itself.

At this point, one easily sees that

(57)
$$I \vee \left[G^{\pm}\right] \vee \left[\left(G^{\pm}\right)^2\right] \vee \left[\left(G^{\pm}\right)^3\right] = \begin{bmatrix} 1 & 1 & 1 & 1 \\ 1 & 1 & 1 & 1 \\ 1 & 1 & 1 & 1 \\ 1 & 1 & 1 & 1 \end{bmatrix},$$

whence

$$(58) \qquad \left[\hat{G}^{\pm}\right] = I \vee \left[G^{\pm}\right] \vee \left[\left(G^{\pm}\right)^{2}\right] \vee \left[\left(G^{\pm}\right)^{3}\right] \vee \cdots = \begin{bmatrix} 1 & 1 & 1 & 1 \\ 1 & 1 & 1 & 1 \\ 1 & 1 & 1 & 1 \\ 1 & 1 & 1 & 1 \end{bmatrix}.$$

This is because once, for a power k, when an (i,j) th entry of $\left[I \cup R \cup R^{2} \cup \cdots \cup R^{k}\right]$ becomes '1', then for all subsequent $n \geq k$, the (i,j) th entry of $\left[I \cup R \cup R^{2} \cup \cdots \cup R^{n}\right]$ will remain '1', whence the (i,j) th entry of the reflexive transitive closure $\hat{R}$ is '1'.

One sees that all entries in each row of $\left[\hat{G}^{\pm}\right]$ are '1's. This says that for each $x \in X$, $\hat{G}^{\pm}(x) = X$; that is, in the undirected graph $G^{\pm}$, there is a path from any vertex to any other vertex. (As explained in Section 3.40, if one vertex has paths to reach all other vertices, then all vertices have paths to reach one another. In terms of the adjacency matrix, if one row of $\left[\hat{G}^{\pm}\right]$ is all '1's, then all rows are all '1's.) One concludes that the graph $G^{\pm}$ is connected (and there is only one connected component, namely, $G^{\pm}$ itself).

Ellipses

5.25 Eightfold Partition   The example in Section 3.30, which attributes the eightfold partition of $P(X \times X)$ by the three subsets R, S, and T, appears thus in adjacency matrix form:

$$(59) \qquad R = \begin{bmatrix} 0 & 1 & 0 & 0 \\ 0 & 0 & 0 & 0 \\ 1 & 0 & 0 & 0 \\ 0 & 0 & 0 & 0 \end{bmatrix}.$$

And here is the Venn diagram with the eight closures R, $R^{=}$, $R^{\pm}$, $\vec{R}$, $\left(R^{\pm}\right)^{=}$, $\left(\vec{R}\right)^{=}$, $\left(\vec{R}\right)^{\pm}$, $R^{=}$:

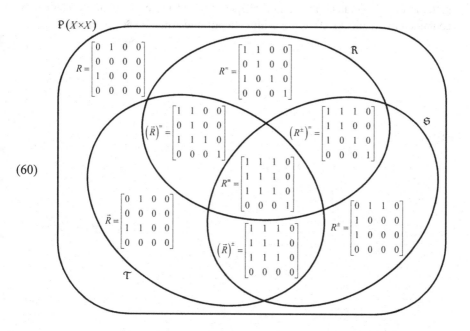

(60)

5.26 Infinite Matrix As a final matrix example, I present the infinite adjacency matrix $[F]$ of the set-valued mapping $F: \mathbb{N} \multimap \mathbb{N}$ from Section 4.17:

$$
(61) \quad [F] = \begin{array}{c} \\ 1 \\ \\ 3 \\ \\ \\ 6 \\ \\ \\ \\ 10 \\ \end{array}
\begin{array}{cccccccccc}
1 & & 3 & & & 6 & & & & 10 \\
\end{array}
\begin{bmatrix}
0 & 0 & 0 & 0 & 0 & 0 & 0 & 0 & 0 & 0 & \cdots \\
0 & 0 & 1 & 0 & 0 & 0 & 0 & 0 & 0 & 0 & \cdots \\
0 & 1 & 0 & 0 & 0 & 0 & 0 & 0 & 0 & 0 & \cdots \\
0 & 0 & 0 & 0 & 1 & 1 & 0 & 0 & 0 & 0 & \cdots \\
0 & 0 & 0 & 1 & 0 & 1 & 0 & 0 & 0 & 0 & \cdots \\
0 & 0 & 0 & 1 & 1 & 0 & 0 & 0 & 0 & 0 & \cdots \\
0 & 0 & 0 & 0 & 0 & 0 & 0 & 1 & 1 & 1 & \cdots \\
0 & 0 & 0 & 0 & 0 & 0 & 1 & 0 & 1 & 1 & \cdots \\
0 & 0 & 0 & 0 & 0 & 0 & 1 & 1 & 0 & 1 & \cdots \\
0 & 0 & 0 & 0 & 0 & 0 & 1 & 1 & 1 & 0 & \cdots \\
\vdots & \vdots & \vdots & \vdots & \vdots & \vdots & \vdots & \vdots & \vdots & \vdots & \ddots \\
\end{bmatrix}.
$$

One easily sees from this that all $F(x)$ and $F^{-1}(x)$ are finite sets, and the patterns of '1's (indicating the nonindependence of the corresponding elements) contained in the blocks $X_n \times X_n$ within the infinite array $\mathbb{N} \times \mathbb{N}$ are evident.

PART II
Functional Entailment

Gegen den Positivismus, welcher bei den Phänomenen stehnbleibt »es gibt nur Tatsachen,« würde ich sagen: nein, gerade Tatsachen gibt es nicht, nur Interpretationen. Wir können kein Faktum »an sich« feststellen: vielleicht ist es ein Unsinn, so etwas zu wollen.

»Es ist alles subjektiv« sagt ihr: aber schon Das ist Auslegung. Das »Subjekt« ist nichts Gegebenes, sondern etwas Hinzu-Erdichtetes, Dahinter-Gestecktes. — Ist es zuletzt nötig, den Interpreten noch hinter die Interpretation zu set zen? Schon Das ist Dichtung, Hypothese.

Soweit Uberhaupt das Wort »Erkenntniss« Sinn hat, ist die Welt erkennbar: aber sie ist anders deutbar, sie hat keinen Sinn hinter sich, sondern unzahlige Sinne. — »Perspektivismus.«

[Against positivism, which halts at phenomena — 'There are only facts' — I would say: No, facts are precisely what there are not, only interpretations. We cannot establish any fact 'in itself': perhaps it is folly to want to do such a thing.

'Everything is subjective,' you say; but even this is interpretation. The 'subject' is not something given, it is something added and invented and projected behind what is there. Finally, is it necessary to posit an interpreter behind the interpretation? Even this is invention, hypothesis.

Insofar as the word 'knowledge' has any meaning, the world is knowable; but it is interpretable otherwise, it has no meaning behind it, but countless meanings. — 'Perspectivism']

— Friedrich Nietzsche (1901)
Kritische Studienausgabe 12:7
Der Wille zur Macht [*The Will to Power*]

The stage is now set for the next movement, in which I shall explicate relational biology in the language of set-valued mappings. In this part, we shall see how the formalism of the general theory of set-valued mappings specializes into functional entailment in a category.

Functional entailment is the entailment of a mapping:

$$\vdash f \, .$$

Robert Rosen considered the biological realization of functional entailment his deepest insight. The innovation is that the processes themselves may be treated like any other material. Rosen arranged the relational organization in his (M,R)-systems in such a way that repair is a mapping that produces as output metabolism that is itself also a mapping: for the 'R' part, instead of just producing an entity on which to operate, it could produce an operator, the 'M' part. The essence of an (M,R)-system is the

'repair ⊢ metabolism'

functional entailment.

A happy happenstance was when Rosen found the connection of this relational theory of biological systems to the algebraic theory of categories, thus equipping himself with a ready-made mathematical tool. While it may not always be immediately apparent, it may be said that almost all of Rosen's scientific works are consequences from a consideration of problems arising in the study of (M,R)-systems. Every time one looks at (M,R)-systems, they have something new to offer.

The definitive exposition on (M,R)-systems remains the now-classic Rosen [1972]. Some of the formal treatments that I undertake in Part II have been suggested in Rosen [1962, 1963].

6
The Logic of Entailment

"I know what you're thinking about," said Tweedledum:
"but it isn't so, nohow."
"Contrariwise," continued Tweedledee, "if it was so, it
might be; and if it were so, it would be; but as it isn't, it ain't.
That's logic."

> — Lewis Carroll (1871)
> *Through the Looking-Glass,*
> *and What Alice Found There*
> Chapter IV Tweedledum and Tweedledee

The defining characteristic of a living system, 'closure to efficient causation', anchors on the key concept of 'functional entailment'. In this chapter, I present a new description in category-theoretic language of this key concept. The emphasis of this alternate description is on the formal setting of the representation.

Truth and Consequences

6.1 The Logic of Entailment   In common usage, the verb *to entail* means 'to have as an inevitable accompaniment' or 'to involve unavoidably'. In logic, the usage is tightened to explicitly involve inference, whence 'to entail' means 'to necessitate as a consequence'.   That the logical usage of 'entailment' in the consequential sense is more stringent than its common usage in the concomitant sense is concisely expressed in the cliché "Correlation does not imply causation."
    When A entails B (whatever entities A and B are), it is denoted

$$(1) \qquad\qquad A \vdash B .$$

The *entailment symbol* $\vdash$ is called 'right tack' in Unicode and has also taken on the ideographic name 'turnstile'. It is used in various branches of mathematics,

A.H. Louie, *The Reflection of Life*, IFSR International Series on Systems Science
and Engineering 29, DOI 10.1007/978-1-4614-6928-5_6,
© Springer Science+Business Media New York 2013

but in all contexts it invariably has the logical meaning that one entity follows another as a necessary consequence.

In the predicate calculus of formal logic, entailment appears in the *conditional statement*

(2) $p \rightarrow q$.

'If p , then q .' The antecedent p whence *entails* the consequent q :

(3) $p \vdash q$,

and this *inferential entailment* is called a *syntactic consequence*. It is independent of any interpretation within the formal system under consideration; that is, logical statement (3) is simply a string of symbols for which the meaning or validity is irrelevant. It is standard in mathematical statements, for clarity, to use the adverb 'then' (in the construction 'If p , then q .') to emphasize the inferential entailment of the consequent, although in common usage, the conjunction 'if', by itself, mostly suffices to define the conditional, and one often simply says 'If p , q .' or 'q if p .'

If the conditional statement (2) is true, then it becomes the *implication*

(4) $p \Rightarrow q$;

' p implies q .' The entailment is then denoted

(5) $p \vDash q$

and is called a *semantic consequence*. This happens when no interpretation within the formal system makes p true and q false (which is the only combination of the truth values of p and q that renders the conditional statement (2) false). When q is a semantic consequence of p , in the parlance of *model theory* in logic, one says " q *models* p ." The *modelling symbol* $\vDash$ is, naturally, also ideographically called 'double turnstile'.

6.2 Infer Versus Imply Incidentally, the difference between statements (2) and (4) (and between statements (3) and (5)) illustrates the contention of *infer* versus *imply* in their logical sense. To infer in an inference (as in statements (2) and (3)) is syntactic; to imply in an implication (as in statements (4) and (5)) is semantic. This distinction agrees with the words' common usage (which is, alas, often mistakenly executed). To infer is to deduce or conclude from facts and reasoning, but there is possibly a certain amount of 'guesswork' (whence feasibly 'error') involved. To imply is to strongly suggest the truth or existence of (an entity that may not be expressly asserted). In communications, the sender of a message implies by putting a suggestion into the message, and the receiver infers by taking a suggestion (not necessarily the same one, whence miscommunication) from the message. The difference between imply and infer is thus in the degree of certainty, the extent of relative truth.

Causality and Inference

6.3 Aristotelian Causation   One may declare that both science and mathematics are, in their different ways, concerned with systems of *entailment*. Aristotelian analysis can be applied to any entailment structure, simply by (as Aristotle did) asking "why?". Statement (1) of 'entailment in the abstract' is realized as causal entailment in the natural world and inferential entailment in the formal world.

Causality is the principle that everything has a cause. *Inference* is the forming of conclusion from premises, the finalization of a proposition as a necessary consequence, in short, 'entailment'. Indeed, the Latin words for entail and infer have the same stem: entail is *adferre* and infer is *inferre* (literally 'to bring to' and 'to bring in', but essentially interchangeable).

It is important to note that I am using the word 'cause' in the *Aristotelian* sense, that is, 'grounds or forms of explanation', *explanatio* (*cf. ML*: 5.2). This includes (but is more expansive than) the common-usage sense of causality as the relation of cause and effect. The realizations of entailment as inference in the formal world and Aristotelian causality in the natural world are exactly analogous, both incarnated in the conditional statement (2). When $p \to q$ and given p, one may formally *infer* q, and naturally q is *enabled*. *Enablement* is the endowment of the means to be or to do, 'making possible', and is verily Aristotle's $\alpha\check{\iota}\tau\iota o\nu$. Our contemporary notion of 'cause', as 'that which produces an effect', has a sense of *fait accompli*. 'Causation' in this modern sense is the act of causing and the act of producing an effect, so it has restrictively evaluated the conditional statement (2) into the implication (4). Aristotle's cause encompasses *potentiality*, which provides the means but not the necessity to (although it may) be 'committed' into the modern 'cause' that is *actuality*.

6.4 Axioms of Relational Biology   In Section 4.15 of *ML*, I wrote (referring to the Natural Law)

> "This equivalence of causality in the natural domain and inference in the formal domain is an epistemological principle, the axiom
>
> *Every process is a mapping.*
>
> Just like the axiom "Everything is a set." leads to the identification of a natural system N and its representation as a set (*cf.* 4.4), mathematical equations representing causal patterns

of natural processes are results of the identification of entailment
arrows and their representations as mappings."

What the two axioms "Everything is a set." and "Every process is a mapping." say
is that we, in the Rashevsky-Rosen school of relational biology, take implicitly as
the mathematical foundation of our science the category **Set** (**Set**-objects are sets
and **Set**-morphisms are mappings; *ML*: A.3). Explicitly, I shall be considering
non-full subcategories **C** of **Set**, in which **C**-objects are sets $A, B, \ldots$, and **C**-hom-
sets $H(A, B)$ are proper subsets of $\mathbf{Set}(A, B) = B^A$ (*ML*: A.7). By extension, we
may also consider *concretizable categories* **C**, those equipped with faithful
functors from **C** to **Set**, for which **C**-objects are sets with structures and **C**-hom-
sets contain morphisms that preserve these structures.

A relational biology based on an interpretation of category axioms within set
theory has been serving our modelling needs. Note the present perfect continuous
tense of the previous sentence, denoting an unbroken past action that continues
right up to the present, an ongoing sequence of events in progress. But it does not
imply that expansions are unwarranted. It is not inconceivable that, in some future
time, relational biology will necessitate a more complicated interpretation of
category axioms, for example, one within hyperset theory.

6.5 Aristotle's Four Causes in a Mapping   We consider entailment the central
concept in relational biology (as it is in the entire scientific enterprise and beyond),
and the Aristotelian analysis of entailment is comprehensively modelled in terms of
its manifestations on a mapping.   The embedding of the four causes as
components of the relational diagram of a mapping is succinctly summarized in
Section 3.1. Without the material cause, the other three causes are related in the
entailment diagram

(6)

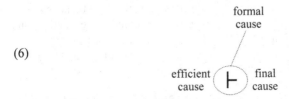

(This is diagram (11) in Chapter 5 of *ML*.)
Material and formal causes are what Aristotle used to explain static things,
that is, things as they are, their *being*, thus the actuality of entailment. Efficient
and final causes are what he used to explain dynamic things, that is, how things
change and come into being, their *becoming*, thus the potentiality of entailment.

6.6 Arrow Differentiation   When a *mapping* $f : A \to B$ is represented in the
element-chasing version $f : a \mapsto b$, its relational diagram in graph-theoretic form
may be drawn as

(7)

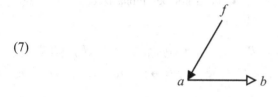

The hollow-headed arrow denotes the *flow* from input $a \in A$ to output $b \in B$, and the solid-headed arrow denotes the induction of or constraint upon this flow by the *processor* f. The processor and output relationship may be characterized 'f *entails* b', which may then be denoted as

(8) $f \vdash b$.

Note that the processor f is *that which entails*, and the output (effect) b is *that which is entailed*. Because of the location of the symbols with respect to the arrows, 'that which entails' may be identified with the (tail of the) solid-headed arrow, and 'that which is entailed' may be identified with the (head of the) hollow-headed arrow. Stated otherwise, if something entails, then it needs to initiate a solid-headed arrow; if something is entailed, then it needs to terminate a hollow-headed arrow.

The final cause (output or effect $b \in B$) of a mapping f is, therefore, entailed by the efficient cause (the processor itself) of f by definition. In other words, 'that which entails' is always the efficient cause, and a process always entails its own output as final cause; whenever there is entailment, final causes are entailed. Thus, the phrases 'final-cause entailment' and 'closure to final causation' are redundancies.

It is when 'that which is entailed' has a *dual role*, an alternate description as some entity in addition to being final cause of its own mapping, that the situation becomes even more interesting. This happens when two processes interact and is modelled by two mappings *in composition*.

Category Theory of Sequential Composition

6.7 Hom-Sets   A category is an interpretation of the category axioms within set theory. Section A.1 of *ML* defines it in terms of hom-sets:

Definition   A *category* **C** consists of:

i.   A collection of *objects*.

ii. For each pair of **C**-objects A, B, a set $\mathbf{C}(A, B)$, the *hom-set* of *morphisms* from A to B. [If $f \in \mathbf{C}(A, B)$, one also writes $f : A \to B$. Often for simplicity, or when the category $\mathbf{C}$ need not be emphasized, the hom-set $\mathbf{C}(A,B)$ may be denoted by $H(A,B)$.]

iii.   For any three objects A , B , C , a *mapping* $\circ : \mathbf{C}(A,B) \times \mathbf{C}(B,C)$ $\rightarrow \mathbf{C}(A,C)$ taking $f : A \rightarrow B$ and $g : B \rightarrow C$ to its *composite* $g \circ f : A \rightarrow C$.

iv.   For each object A , there exists a morphism $1_A \in \mathbf{C}(A,A)$, called the *identity morphism* on A .

These entities satisfy the following three axioms:

(c1)   *Uniqueness*: $\mathbf{C}(A,B) \cap \mathbf{C}(C,D) = \varnothing$ unless $A = C$ and $B = D$.   [Thus each morphism $f : A \rightarrow B$ uniquely determines its *domain* $A = \mathrm{dom}(f)$ and *codomain* $B = \mathrm{cod}(f)$: different hom-sets are mutually exclusive.]

(c2)   *Associativity*:   If $f : A \rightarrow B$, $g : B \rightarrow C$, $h : C \rightarrow D$, so that both $h \circ (g \circ f)$ and $(h \circ g) \circ f$ are defined, then $h \circ (g \circ f) = (h \circ g) \circ f$.

(c3)   *Identity*:   For each object A , the identity morphism on A , $1_A : A \rightarrow A$, has the property that for any $f : A \rightarrow B$, $f \circ 1_A = f$, and for any $g : C \rightarrow A$, $1_A \circ g = g$ [which leads demonstrably to the uniqueness of 1_A in $\mathbf{C}(A,A)$].

If one denotes the collection of all $\mathbf{C}$-objects as $\mathbf{OC}$ and the collection of all $\mathbf{C}$-morphisms (*arrows*) as $\mathfrak{A}\mathbf{C}$, then one has

$$(9) \qquad\qquad \mathfrak{A}\mathbf{C} = \bigcup_{A,B \in \mathbf{OC}} \mathbf{C}(A,B).$$

Axiom (c1) serves to synthesize $\mathfrak{A}\mathbf{C}$ as the *union of pairwise disjoint hom-sets*.

Item ii above postulates the existence of a mapping H that assigns to each ordered pair of objects $(A,B) \in \mathbf{OC} \times \mathbf{OC}$ a hom-set $H(A,B) \subset \mathfrak{A}\mathbf{C}$, that is,

$$(10) \qquad\qquad H : \mathbf{OC} \times \mathbf{OC} \rightarrow \mathrm{P}(\mathfrak{A}\mathbf{C}),$$

or, equivalently (and better, now that we are equipped with the nuances of the theory), the set-valued mapping

$$(11) \qquad\qquad H : \mathbf{OC} \times \mathbf{OC} \multimap \mathfrak{A}\mathbf{C} .$$

Axiom (c1), then, says this set-valued mapping H is *injective* (*cf.* Definition 2.7.iii).

6.8 The Collection of Arrows   Alternatively, one may define a category in terms of arrows thus:

Definition   A *category* C consists of:

  i'. A collection OC of *objects*

 ii'. A collection $\mathcal{A}C$ of *arrows* (*morphisms*), equipped with two mappings dom and cod:

(12)
$$\begin{cases} \text{dom} : \mathcal{A}C \to OC \\ \text{cod} : \mathcal{A}C \to OC \end{cases}$$

iii'. A (*sequential*) *composition* mapping

(13)
$$\circ : \mathcal{A}C \times_{OC} \mathcal{A}C \to \mathcal{A}C$$

(where the domain

(14)
$$\mathcal{A}C \times_{OC} \mathcal{A}C = \left\{ (f,g) \in \mathcal{A}C \times \mathcal{A}C : \text{dom}(g) = \text{cod}(f) \right\}$$

is a proper subset of $\mathcal{A}C \times \mathcal{A}C$, called the '*product over* OC', and an ordered pair $(f,g) \in \mathcal{A}C \times_{OC} \mathcal{A}C$ is called a '*composable pair of morphisms*'), taking (f,g) to its *composite* $g \circ f$, such that

(15)
$$\text{dom}(g \circ f) = \text{dom}(f) \quad \text{and} \quad \text{cod}(g \circ f) = \text{cod}(g)$$

iv'. A mapping

(16)
$$\text{id} : OC \to \mathcal{A}C$$

that sends a C-object A to the *identity morphism* $\text{id}(A) = 1_A$ on A, such that

(17)
$$\text{dom}(1_A) = \text{cod}(1_A) = A$$

These entities satisfy the following two axioms:

($c2'$) *Associativity*: If $(f,g) \in \mathfrak{R}C \times_{\mathbf{oc}} \mathfrak{R}C$ and $(g,h) \in \mathfrak{R}C \times_{\mathbf{oc}} \mathfrak{R}C$, so that both $h \circ (g \circ f)$ and $(h \circ g) \circ f$ are defined, then $h \circ (g \circ f) = (h \circ g) \circ f$.

($c3'$) *Identity*: For any $f : A \to B$, $g : C \to A$, one has $f \circ 1_A = f$, $1_A \circ g = g$.

The hom-sets may be defined as

(18) $\mathbf{C}(A,B) = \{ f \in \mathfrak{R}C : \operatorname{dom}(f) = A,\ \operatorname{cod}(f) = B \}$.

Then, one again has a mapping H that assigns to each ordered pair of objects $(A,B) \in \mathbf{OC} \times \mathbf{OC}$ the hom-set $H(A,B) = \mathbf{C}(A,B) \subset \mathfrak{R}C$, that is,

(19) $H : \mathbf{OC} \times \mathbf{OC} \to \mathbf{P}(\mathfrak{R}C)$,

and the equivalent set-valued mapping

(20) $H : \mathbf{OC} \times \mathbf{OC} \multimap \mathfrak{R}C$.

6.9 Partitioning of $\mathfrak{R}C$ Note that in the arrow-based definition 6.8, a statement ($c1'$) corresponding to 6.7($c1$) on the mutual exclusiveness of hom-sets is not required as a separate axiom. Since dom and cod are postulated in ii$'$ as mappings, their Cartesian product set-valued mapping (Definition 2.28.iii)

(21) $\operatorname{dom} \times \operatorname{cod} : \mathfrak{R}C \multimap \mathbf{OC} \times \mathbf{OC}$

defined by

(22) $\operatorname{dom} \times \operatorname{cod}(f) = \{ (\operatorname{dom}(f), \operatorname{cod}(f)) \}$

is single-valued. So by Theorem 2.15.i, its inverse set-valued mapping is injective. But $(\operatorname{dom} \times \operatorname{cod})^{-1}$ is by definition

(23) $(\operatorname{dom} \times \operatorname{cod})^{-1}(A,B) = \{ f \in \mathfrak{R}C : \operatorname{dom}(f) = A,\ \operatorname{cod}(f) = B \}$,

that is, the set-valued mapping that sends a pair of **C**-objects A, B to the hom-set $\mathbf{C}(A,B)$ of morphisms from A to B, which of course is

(24) $(\operatorname{dom} \times \operatorname{cod})^{-1} = H : \mathbf{OC} \times \mathbf{OC} \multimap \mathfrak{R}C$.

The set-valued mapping H being injective dictates precisely that

$(c1')$ if $(A,B) \neq (C,D)$ in $\mathbf{OC} \times \mathbf{OC}$, then $\mathbf{C}(A,B) \cap \mathbf{C}(C,D) = \varnothing$

(*cf.* Definition 2.7.iii). Stated otherwise, a statement $(c1')$ on the mutual exclusiveness of hom-sets is already entailed by postulate ii'. In this formulation, $\mathfrak{AC}$ is analytically *partitioned into mutually disjoint hom-sets*.

6.10 Composition as Mapping   One sees from Definitions 6.7.iii and 6.8.iii' that (sequential) composition is an integral part of the definition of a category. Composition ∘ may be considered a *binary operation* on $\mathfrak{AC}$, one that takes a pair of composable **C**-morphisms to a **C**-morphism:

(25) $\circ : (f,g) \mapsto g \circ f$.

It is important to note that the binary operation ∘ is not defined for all pairs of **C**-morphisms: the codomain of the first argument f must be identical to the domain of the second argument g for the binary operation to proceed. Indeed, the domain of ∘ is not the Cartesian product $\mathfrak{AC} \times \mathfrak{AC}$ but the set $\mathfrak{AC} \times_{\mathbf{OC}} \mathfrak{AC}$ (the 'product over $\mathbf{OC}$' in (14)), which is a subset of $\mathfrak{AC} \times \mathfrak{AC}$ restricted by the very requirement $\mathrm{dom}(g) = \mathrm{cod}(f)$. The set $\mathrm{dom}(\circ)$, containing pairs of composable **C**-morphisms, is a union of Cartesian product of hom-sets over pairs that share a 'common middle':

(26) $\mathrm{dom}(\circ) = \mathfrak{AC} \times_{\mathbf{OC}} \mathfrak{AC} = \bigcup_{A,B,X \in \mathbf{OC}} \mathbf{C}(A,X) \times \mathbf{C}(X,B)$.

The codomain of ∘ is the collection of arrows $\mathfrak{AC}$, the union of all the mutually disjoint **C**-hom-sets:

(27) $\mathrm{cod}(\circ) = \mathfrak{AC} = \bigcup_{A,B \in \mathbf{OC}} \mathbf{C}(A,B)$.

Although the **C**-morphisms may not necessarily be mappings, composition itself is a mapping (from a set to a set). This is because, for each pair of **C**-objects A and B, the hom-set of **C**-morphisms $\mathbf{C}(A,B)$ is above all a set, in any category **C** (whether concretizable or not; roughly, whether it is a category of 'sets with structure' or not). Thus, the category **Set** involves itself in an essential way in every category.

When $\mathbf{C} = \mathbf{Set}$ and when sequential composition of the composable pair $(f,g) \in H(A,X) \times H(X,B)$ is presented as an element-chasing relational diagram (where, naturally, $a \in A$, $b \in B$, and $x \in X$), one has

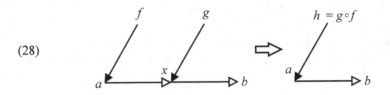

(28)

with corresponding entailment diagram

(29) $f \vdash x, \quad g \vdash b \quad \Rightarrow \quad h = g \circ f \vdash b.$

The sequential composite $h = g \circ f$ entails the (final) final cause b. The 'common middle' element $x \in X$ is entailed as the final cause of its own processor f, and it is also relayed as the material cause of the next processor g; the final cause of f is the material cause of g. It is this second role of the 'relayed element' $x \in X$ that is remarkable (in the sense of 'worthy of a remark', i.e., 'worth naming') in sequential composition, whence called *material entailment*.

6.11 Composition as Relay One may note that in the composition $g \circ f$, the equality condition

(30) $\operatorname{cod}(f) = \operatorname{dom}(g)$

is sufficient but not strictly necessary. All that is required is the inclusion

(31) $\operatorname{ran}(f) \subset \operatorname{dom}(g)$

(which is implied by (30), since $\operatorname{ran}(f) \subset \operatorname{cod}(f)$). This is because if $x \in \operatorname{cod}(f) \sim \operatorname{ran}(f)$, then $x \neq f(a)$ for any $a \in A$, and therefore, x would not be involved in the 'relay' $g \circ f : a \mapsto f(a) \mapsto g(f(a))$, whence it is not necessary to insist that $x \in \operatorname{dom}(g)$.

Recall (Section 5.5) that, in terms of adjacency matrices,

(32) $[g \circ f] = [f][g],$

and the (k, j) th entry in the matrix $[g \circ f]$ of the sequential composite is

(33) $(g \circ f)_{kj} = \sum_{i=1}^{m} f_{ki}\, g_{ij}.$

If $x_i \notin \mathrm{ran}(f)$, then the i th column of the adjacency matrix $[f]$ consists entirely of '0's; that is, $f_{ki} = 0$ for all k. This means $f_{ki}\, g_{ij} = 0$ for all k and j, whatever values g_{ij} take. In other words, in calculating the entries (33), one does not have to know the entries of the i th row of the adjacency matrix $[g]$; that is, it is not necessary to insist that one has $x_i \in \mathrm{dom}(g)$ whence a definition for $g(x_i)$.

It is, nevertheless, conventional to impose the equality (30), which may be achieved (when (31) is satisfied) by appropriate 'readjustment' of $X = \mathrm{cod}(f)$ to match $\mathrm{dom}(g)$: for all sets $X \in \mathcal{OC}$ such that $\mathrm{ran}(f) \subset X$, one may *define* $X = \mathrm{cod}(f)$ and so $f \in H(A,X) \subset \mathcal{AC}$.

Category Theory of Hierarchical Composition

6.12 Exponential   If a category $\mathbf{C}$ has (binary) products, then there is an induced functor $\bullet \times \bullet : \mathbf{C} \times \mathbf{C} \to \mathbf{C}$. In particular, fixing a $\mathbf{C}$-object Y, consider the induced functor $\bullet \times Y : \mathbf{C} \to \mathbf{C}$ that maps $\mathbf{C}$-objects $X \mapsto X \times Y$. Then, by definition (*ML*: A.48), this functor $\bullet \times Y : \mathbf{C} \to \mathbf{C}$ has a *right adjoint* $G : \mathbf{C} \to \mathbf{C}$ (dependent on Y) if and only if one has the natural isomorphism $\varphi : \mathbf{C}(X \times Y, Z) \cong \mathbf{C}(X, GZ)$. If one denotes this right adjoint $G(\,\bullet\,) = (\,\bullet\,)^Y$, then

(34) $$\mathbf{C}(X \times Y, Z) \cong \mathbf{C}(X, Z^Y).$$

The $\mathbf{C}$-object Z^Y is called an *exponential* (*ML*: A.52). Specifying the adjunction $\langle\, \bullet \times Y, (\,\bullet\,)^Y, \varphi \,\rangle : \mathbf{C} \to \mathbf{C}$ amounts to assigning to each pair of $\mathbf{C}$-objects X and Z the $\mathbf{C}$-morphism $e : Z^X \times X \to Z$, called the *evaluation morphism*, which is natural in Z and universal from $\bullet \times X$ to Z. (*Cf. ML*: A.53 on Cartesian closed category.)

In the category **Set**:

i. The exponential Z^Y is the hom-set $\mathbf{Set}(Y,Z)$ of all mappings from Y to Z.

ii. $Z^Y \times Y^X \to Z^X$ is a natural transformation, which agrees with (sequential) composition of mappings (*cf.* definition (9)).

iii. The evaluation mapping $e : Z^X \times X \to Z$ is $e : (g,x) \mapsto g(x)$ for $g : X \to Z$ and $x \in X$.

Note that, in a general category, given **C**-objects Z and Y, the exponential Z^Y does not necessarily exist. Even when it does exist, it is simply a **C**-object; it does not necessarily correspond to the hom-set $C(Y,Z)$. This is (roughly) because there is not always a 'good' way to define a **C**-structure on the hom-set $C(Y,Z)$ to turn it into a **C**-object.

6.13 Hierarchical Composition   In a category **C** in which exponentials exist and they correspond to hom-sets (an example is, of course, **Set**), a composition that is different in kind from sequential composition may be defined. This is called *hierarchical composition*, embodied in the natural isomorphism:

$$(35) \qquad\qquad C\left(X, B^A\right) \cong H\left(X, H(A,B)\right),$$

through which a **C**-morphism $g \in C\left(X, B^A\right)$ may be interpreted as a mapping (i.e., **Set**-morphism) $g \in H\left(X, H(A,B)\right)$ (where I have used the hom-set notation $H(Y,Z)$ for a subset of the set $\mathbf{Set}(Y,Z) = Z^Y$ of all mappings from set Y to set Z).

The codomain $H(A,B)$ of $g \in H\left(X, H(A,B)\right)$ is a collection of mappings: for $x \in X$, $f = g(x) : A \to B$. Since the codomain of g contains f (i.e., $f \in H(A,B)$), the mapping g may be considered to occupy a higher *hierarchical level* than the mapping f. Let the element chases be $f : a \mapsto b$ and $g : x \mapsto f$. Then one has the relational diagram

(36)

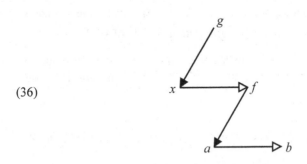

with the corresponding composition of entailment diagrams

$$(37) \qquad\qquad g \vdash f, \quad f \vdash b \quad \Rightarrow \quad g \vdash f \vdash b.$$

The 'common middle' here is the mapping $f \in H(A,B)$: the final cause of g is the efficient cause of f .

6.14 Functional Entailment   What is relayed in a hierarchical composition, that which the first mapping entails, is now the processor of the next mapping; that is, the dual role of the first final cause is that of the next efficient cause. This different mode of entailment, the *entailment of an efficient cause*, is given the name of *functional entailment* and may be denoted succinctly

(38) $\vdash f$.

Because of the natural isomorphism

(39) $H(X \times A, B) \cong H(X, H(A,B))$

(consequence of (34) and (35)), a mapping $g \in H(X, H(A,B))$ that entails another mapping and has a hom-set as codomain may be considered equivalently as the isomorphic $g \in H(X \times A, B)$ that has a simple set as codomain. Thus, one sees that, while hierarchical composition is *formally* different from sequential composition (as is evident from a comparison of the entailment chains (29) and (37)), functional entailment is not *categorically* different. This isomorphic resolution (39) of the entailment of an efficient cause has important biological explications, as we shall see.

7

The Imminence Mapping

For as in one body we have many members, and not all the
members have the same function.

— *Epistola Beati Pauli Apostoli ad Romanos* 12:4

Clef Systems and (M,R)-Systems

Division of labour is a crucial characteristic of living systems. In (M,R)-systems, the class of formal systems that are the defining relational models of organisms, the internal processes are functionally divided into two types of components, *metabolism* (that which materially entails) and *repair* (that which functionally entails). The reader is referred to Chapter 11 of *ML* for a review.

The modelling relation (*cf. ML*: Chapter 4) encodes processes in natural systems as mappings in formal systems. This establishes a correspondence between causal entailment in the natural domain and inferential entailment in the formal domain. In particular, efficient causes in natural systems are encoded as efficient causes of mappings.

A hierarchical cycle (*cf. ML*: 6.16–6.17) is, then, the formal system representation of a closed path of efficient causation in a natural system. So, trivially, one has the following:

7.1 Lemma (*ML*: 6.19) *A natural system has a model containing a hierarchical cycle if and only if it has a closed path of efficient causation.*

A natural system is closed to efficient causation if its every efficient cause is entailed within the system (*ML*: 6.23). Thus, equivalently,

7.2 Theorem (*ML*: 6.25) *A natural system is closed to efficient causation if and only if it has a model in which all efficient causes are involved in hierarchical cycles.*

A.H. Louie, *The Reflection of Life*, IFSR International Series on Systems Science
and Engineering 29, DOI 10.1007/978-1-4614-6928-5_7,
© Springer Science+Business Media New York 2013

The equivalence also allows the description *closed to efficient causation* to be used on formal systems, those with all efficient causes involved in hierarchical cycles.

Instead of the verbose 'closed-to-efficient-cause system' or 'system that is closed to efficient causation', in Louie and Poli [2011] we have introduced a new term '*clef* system' (for *cl*osed to *ef*ficient causation) with the

7.3 Definition   A natural system is *clef* if and only if it has a model that has all its processes contained in hierarchical cycles.

The word 'clef' means 'key'; so this terminology has the added bonus of describing the importance of the class of *clef systems*. Analogously, a *clef* formal system is one that has all its mappings contained in hierarchical cycles.

7.4 Organism   The answer to the "What is life?" question according to the Rashevsky-Rosen school of relational biology is, tersely, that an *organism*—the term is used in the sense of an 'autonomous life form', i.e. any living system (including, in particular, cells)—admits a specific kind of relational description, that it is 'closed to efficient causation'. Explicitly, a material system is an organism if and only if its *every efficient cause is functionally entailed* within the system (*ML*: 11.29). Stated otherwise, an organism is a clef system. This 'self-sufficiency' in efficient causation, this closure to functional entailment, is what we implicitly recognize as the one feature that distinguishes a living system from a nonliving one.

In relation to (M,R)-systems (*cf. ML*: Chapters 11 and 12), we may state the Postulate of Life: A natural system is an organism if and only if it realizes an (M,R)-system (*ML*: 11.28). Thus, an (M,R)-system is the very model of life, and conversely, life is the very realization of an (M,R)-system.

7.5 (M,R)-System Maps   The entailment patterns of the three (M,R)-system maps are as follows:

(1)   Metabolism $f \vdash b$

(2)   Repair $\Phi \vdash f$

(3)   Replication $b \vdash \Phi$

whence they functionally entail one another in cyclic permutation in the entailment diagram:

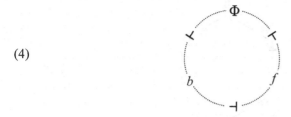

(4)

The three maps $\{f, \Phi, b\}$ appear in a hierarchical cycle, thus:

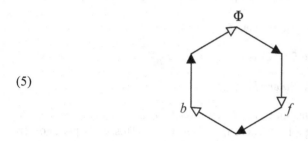

(5)

One sees that there is cyclic permutational symmetry among the three maps $\{f, \Phi, b\}$: the three functionally entailing processes ⟨metabolism, repair, replication⟩ may be any one of $\langle f, \Phi, b \rangle$, $\langle \Phi, b, f \rangle$, or $\langle b, f, \Phi \rangle$. Indeed, this 'multitasking' of components has been observed in cells: metabolites may take on epigenetic functions, enzymes may themselves be metabolized, etc.

7.6 Unfolding The relational diagram of the simplest (M,R)-system is

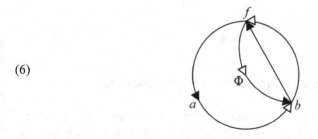

(6)

which unfolds into the hierarchical cycle

(7)

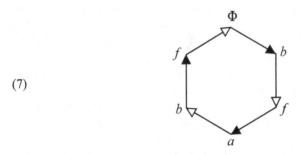

Note that f has three distinct roles in this system:

i. In $f : a \mapsto b$, f is the *efficient* cause.

ii. In $\Phi : b \mapsto f$, f is the *final* cause.

iii. In $b : f \mapsto \Phi$, f is the *material* cause.

A corresponding statement may be made about b and, indeed, when an isomorphism $a \cong \Phi$ is properly defined, likewise for Φ. This is simply another way of illustrating the cyclic permutational symmetry among the three maps $\{f, \Phi, b\}$.

For each of the other two alternate encodings of replication, the (M,R)-system also unfolds into a similar hierarchical cycle

(8)

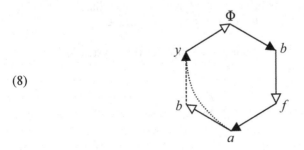

with an appropriate choice of y, the material cause of replication.

7.7 Process Closure   Let me emphasize that the defining property of an (M,R)-system is that of a clef system of mutually functionally entailing mappings. In its simplest form, it is three maps $\{f, \Phi, b\}$ in a hierarchical cycle as in (5), but the maps actually represent general processes of a living system, and {metabolism, repair, replication} is only one possible realization.

Recall (*ML*: 11.34) that these processes describe relational, functional organizations, and there is no one-to-one correspondence between them and the structures that realize them. A functional organization cuts across physical structures, and a physical structure is simultaneously involved in a variety of functional activities. An (M,R)-system is not realized by identifying its objects and mappings in 'concrete' biological components and processes. To tackle the biological realization problem of (M,R)-systems, one ought not to be seeking physicochemical implementations of what the relations *are*, but ought instead to be seeking interpretations of what the relations *do*.

The closure in efficient causation of an (M,R)-system is that of self-sufficiency in functional entailment; in short, clef is a *process closure*. A common interpretative error is the misleading assignment of the label 'metabolic closure' or 'catalytic closure' to (M,R)-systems. Even in the canonical {metabolism, repair, replication} realization, enzyme-catalyzed metabolism is represented only by $f : a \mapsto b$, one of the three mappings (1)–(3) in the simplest set $\{f, \Phi, b\}$. So the 'metabolic/catalytic closure' label offers an incomplete description at best.

(M,R)-Networks

7.8 Definition (*ML*: 7.1)   A *formal system* is a pair $\langle S, F \rangle$, where S is a set and F is a collection of observables of S, *i.e.* $F \subset \bullet^S$, such that $0 \in F$, where 0 is (the equivalence class of) the constant mapping on S.

The category of formal systems is the subject of Chapter 7 of *ML*. Recall $0 \in F$ is an algebraic requirement; the role of 0 is to identify the set S itself and to define the property of 'belonging to S'. Indeed, for each $f \in F$, $\mathrm{dom}(f) = S$; so the collection F of observables implicitly defines the set S, whence a formal system may alternatively be defined as a collection F of observables on a common domain.

7.9 Definition (*ML*: 7.17)   A *model* of a natural system N is a finite collection of formal systems $\{\langle S_i, F_i \rangle : i = 1, ..., n\}$ such that the collections of mappings $\{F_i : i = 1, ..., n\}$ satisfy the entailment requirements of the modelling relation (*cf. ML*: Chapter 4). Each formal system $\langle S_i, F_i \rangle$ is called a *component* of the model.

The collection of all models of a natural system N is denoted $\mathbf{C}(N)$ (*ML*: 7.27). $\mathbf{C}(N)$ is a lattice (*ML*: 7.28) as well as a category (*ML*: 7.29).

An (M,R)-network that models a natural system N is a member of $\mathbf{C}(N)$. An (M,R)-network is an entailment network of a finite collection of metabolism and repair components, with the formal definition:

7.10 Definition (*ML*: 13.2)   *Metabolism* and *repair* are input-output systems that are collected as *components* into a network. They are formal systems with the following further category-theoretic structures:

(*a*)   A *metabolism component* is a formal system $M_i = \langle A_i, H(A_i, B_i) \rangle$.

(*b*)   A *repair component* is a formal system $R_i = \langle Y_i, H(Y_i, H(A_i, B_i)) \rangle$.

(*c*)   A *metabolism-repair network*, i.e. an *(M,R)-network*, is a finite collection of pairs of metabolism and repair components $\{(M_i, R_i) : i \in I\}$, connected in a model network. In particular, the outputs of a repair component R_i are observables in $H(A_i, B_i)$ of its corresponding metabolism component M_i. The metabolism components may be connected among themselves by their inputs and outputs (i.e. by $B_k \subset A_j$ for some $j, k \in I$). Repair components must receive at least one input from the outputs of the metabolism components of the network (i.e. $Y_i = \prod_{k=1}^{n_i} B_{i_k}$ with $n_i \geq 1$ and where each $i_k \in I$).

The connections specified in (*c*) are the requisite ones; an (M,R)-network may have additional interconnections among its components and with its environment.

Let N be a natural system and let $\kappa(N)$ be the collection of all efficient causes in N. (A standard Aristotelian term in Greek for 'efficient cause' is ὅθεν ἡ ἀρχὴ τῆς κινήσεως, where κινήσεως (*kinēseōs*) = motion; whence the symbol κ.) An (M,R)-network that models N may be denoted $\varepsilon(N) \in \mathbf{C}(N)$, where ε is an encoding functor (*ML*: 4.14). Efficient causes of causal entailment are correspondingly encoded into mappings of inferential entailment; thus, the collection of all efficient causes (i.e. mappings) in the (M,R)-network $\varepsilon(N)$ may be represented by the encoding $\varepsilon[\kappa(N)]$. Then the requisite metabolism and repair mappings in $\varepsilon(N)$ specified in (*c*) are such that, for all $i \in I$,

(9) $H(A_i, B_i) \subset \varepsilon[\kappa(N)]$ and $H(Y_i, H(A_i, B_i)) \subset \varepsilon[\kappa(N)]$.

Additional interconnections among the components of N and between N and its environment are processes represented as mappings in $\varepsilon(N)$, so these mappings are also in $\varepsilon\left[\kappa(N)\right]$. One, therefore, sees that an (M,R)-network $\varepsilon(N)$ is completely defined by its collection $\varepsilon\left[\kappa(N)\right]$ of efficient causes.

When only one encoding of a natural system N into an (M,R)-network model $\varepsilon(N)$ is under consideration, I shall, for notational simplicity, drop the encoding symbol ε and use N to denote both the natural system and its (M,R)-network model. Thus, 'an (M,R)-network $\varepsilon(N)$ that models a natural system N ' abbreviates to 'an (M,R)-network N '. Likewise the symbol $\kappa(N)$ shall denote the collection of efficient causes in both the natural system and the formal system.

7.11 Definition   A *metabolism-repair system*, i.e. an *(M,R)-system*, is an (M,R)-network that is closed to efficient causation.

An (M,R)-system is, in particular, a clef system (*cf.* Definition 7.4).

7.12 Immanent Causation   The process closure that is the closure in efficient causation of an (M,R)-system N means every process is functionally entailed within N. For every process $\alpha \in \kappa(N)$ in N, there exists another process $\beta \in \kappa(N)$ such that $\beta \vdash \alpha$. Symbolically (*cf.* ML: 6.23),

(10) $$\forall \alpha \in \kappa(N) \; \exists \beta \in \kappa(N) : \; \beta \vdash \alpha.$$

Entailment, itself, entails the existence of an efficient cause. If a morphism $\alpha \in H(Y,Z)$ is functionally entailed, then there exists a morphism $\beta \in H(X, H(Y,Z))$ (which implicitly implies the existence of a set X and an element $x \in X$) such that $\beta : x \mapsto \alpha$. The entailment of the *existence* of an entity (often on a higher hierarchical level) is *immanent causation* (*ML*: 5.18) and may be summarized as

(11) $$\left(\vdash \alpha \right) \vdash \left(\exists \beta : \; \beta \vdash \alpha \right).$$

7.13 That Which Functionally Entails   In particular, if a process $\alpha \in \kappa(N)$ in an (M,R)-system N is a metabolism map $\alpha : A_i \to B_i$, then it is entailed by a repair map

(12) $$\beta : Y_i \to H\left(A_i, B_i \right),$$

and if $\alpha : Y_i \to H\left(A_i, B_i \right)$ is a repair map, then it is entailed by a replication map

(13) $$\beta : Z_i \rightarrow H\left(Y_i, H\left(A_i, B_i\right)\right),$$

where the domains Y_i, Z_i are Cartesian products of sets within the (M,R)-system. 'That which functionally entails', as one sees from (12) and (13), has the form $\beta : X \rightarrow H(Y,Z)$; i.e.

(14) $$\beta \in H\left(X, H\left(Y, Z\right)\right).$$

Thus, the stringent clef requirement of 'every efficient cause entailed within the system' hinges on the availability of maps in hom-sets

(15) $$H\left(X, H\left(Y, Z\right)\right)$$

in our chosen universe of models that contains the (M,R)-systems.

Let N be a natural system. In the category **Set**, without restriction, models of N are built on **Set**-objects (sets) and **Set**-morphisms (mappings). Recall (Section 6.4), however, that in general our models are drawn from smaller non-full subcategories **C** of **Set**, in which **C**-objects are a selection of sets $A, B, ...$, and **C**-hom-sets $H(A,B)$ are proper subsets of $\mathbf{Set}(A,B) = B^A$. Within this chosen categorical universe of models, one has the following:

7.14 Definition   A *model of N in the category* **C** is a member of $\mathbf{C}(N)$ with **C**-objects and **C**-morphisms.

Let (as introduced in Sections 6.7 and 6.8) $\mathcal{O}\mathbf{C}$ be the collection of **C**-objects (that are sets) and $\mathcal{A}\mathbf{C}$ be its collection of **C**-morphisms (that are mappings). If a model in the category **C** consists of the finite collection of formal systems $\left\{\langle S_i, F_i \rangle : i = 1,...,n\right\}$, then for all $i = 1,...,n$,

(16) $$S_i \in \mathcal{O}\mathbf{C} \quad \text{and} \quad F_i \subset \mathcal{A}\mathbf{C}.$$

A model of N in the category **C** may alternatively be described as a formal system that is a network of mappings in $\mathcal{A}\mathbf{C}$, in which case one may alternately refer to 'a system N in the category **C**' when its collection $\kappa(N)$ of efficient causes is a subset of **C**-morphisms:

(17) $$\kappa(N) \subset \mathcal{A}\mathbf{C}.$$

It is appropriate here to recall that a generic natural system N is such that its lattice $\mathbf{C}(N)$ of models does not have a greatest element (cf. ML: 8.24). The nongeneric collection of natural systems that do have largest models is that of the simple systems (cf. ML: Chapter 8), and the complementary collection of natural systems that are not simple is that of the complex systems (cf. ML: Chapter 9). The no-largest-model property entails the restriction that a model of a complex system N must not contain all the observables of N, that the containment (17) must be proper.

Entailment in ℛC

7.15 Range of a Mapping   For a mapping $f \in \text{ℛC}$, let $\text{Imm}(f)$ be the collection of all **C**-morphisms that lie in the range of f , i.e. the subset

$$(18) \qquad\qquad \text{Imm}(f) = \text{ℛC} \cap \text{ran}(f)$$

of ℛC . This defines a mapping

$$(19) \qquad\qquad \text{Imm} : \text{ℛC} \to \text{P}(\text{ℛC}),$$

which is, of course, equivalently a set-valued mapping

$$(20) \qquad\qquad \text{Imm} : \text{ℛC} \multimap \text{ℛC},$$

i.e. Imm is a relation on ℛC and a member of the monoid $\mathbf{Rel}(\text{ℛC}, \text{ℛC})$ (cf. Section 3.21).

In Section 5.12 of ML, I explicated in detail the impossibility of the existence of the nontrivial self-entailed mapping, the *ouroboros*

$$(21) \qquad\qquad f \vdash f$$

in naive set theory (and categories of sets with structure, which form the universe with which we are concerned). The absence of self-entailment (21) means for every $f \in \text{ℛC}$

$$(22) \qquad\qquad f \notin \text{ran}(f),$$

whence

$$(23) \qquad\qquad f \notin \text{Imm}(f),$$

or equivalently,

$$(24) \qquad \qquad \mathrm{Imm}(f) \subset \mathfrak{R}\mathbf{C} \sim \{f\}.$$

That is to say, every $f \in \mathfrak{R}\mathbf{C}$ is independent of itself (*cf.* Section 4.6) under the set-valued mapping $\mathrm{Imm}: \mathfrak{R}\mathbf{C} \multimap \mathfrak{R}\mathbf{C}$. Note that this self-independence of elements is a prerequisite condition for the cardinality theorems in Chapter 4.

7.16 Imminence   It is evident from definition (18) of the set-valued mapping $\mathrm{Imm}: \mathfrak{R}\mathbf{C} \multimap \mathfrak{R}\mathbf{C}$ that for $f, g \in \mathfrak{R}\mathbf{C}$,

$$(25) \qquad \qquad g \in \mathrm{Imm}(f) \quad \text{iff} \quad f \vdash g,$$

i.e. iff f functionally entails g. This equivalence makes the set-valued mapping Imm crucially important in our study of functional entailment. One may note that $\mathrm{Imm}(f) = \varnothing$ iff $f \in \mathfrak{R}\mathbf{C}$ functionally entails no mappings in $\mathfrak{R}\mathbf{C}$, whence the core of $\varnothing$ by Imm (Definition 2.16.ii), the set $\mathrm{Imm}^{+1}(\varnothing) \subset \mathfrak{R}\mathbf{C}$, contains all the C-morphisms that do not functionally entail others.

By definition (*cf.* Definition 2.4), $\mathrm{Imm}(f) \neq \varnothing$ iff $f \in \mathrm{cor}(\mathrm{Imm})$. A nonempty set $\mathrm{Imm}(f)$, being the collection of all C-morphisms that lie in the range of f, is the collection of all the f-entailed entities that can themselves entail. Members of $\mathrm{Imm}(f)$ are efficient causes; if $g \in \mathrm{Imm}(f)$, then the further action of g is *imminent*. Symbolically, one has

$$(26) \qquad \qquad (\exists g \in \mathfrak{R}\mathbf{C}: f \vdash g) \vdash (g \vdash).$$

(Compare this with immanent causation in (11) above; for a discussion of *immanence* versus *imminence*, see *ML*: 10.3.) In other words, the set $\mathrm{Imm}(f)$ contains all possible further actions arising from f (whence also entails *anticipation*; *cf.* Axiom of Anticipation, *ML*: 10.4), which we may consider the *imminence* of f, the 'global' manifestation of the 'local' functional entailment. I shall, therefore, give the set-valued mapping $\mathrm{Imm}: \mathfrak{R}\mathbf{C} \multimap \mathfrak{R}\mathbf{C}$ defined in (18) the natural name of *imminence mapping* (which explains the use of the expression 'Imm' as the symbol for this set-valued mapping). This is a key concept of this monograph (indeed, the mathematical formalism contained in the Pentateuchus of Part I is presented to lead up to this special set-valued mapping), so I shall restate its definition explicitly:

7.17 Definition   Let **C** be a subcategory of **Set**, and let $\mathfrak{A}\mathbf{C}$ be its collection of morphisms. The *imminence mapping of the category* **C** (also the *imminence mapping on* $\mathfrak{A}\mathbf{C}$) is the set-valued mapping

$$(27) \qquad\qquad \operatorname{Imm}: \mathfrak{A}\mathbf{C} \multimap \mathfrak{A}\mathbf{C}$$

(also $\operatorname{Imm}_{\mathbf{C}} : \mathfrak{A}\mathbf{C} \multimap \mathfrak{A}\mathbf{C}$ when the dependence of Imm on the category **C** needs to be emphasized) defined, for each mapping $f \in \mathfrak{A}\mathbf{C}$, by

$$(28) \qquad\qquad \operatorname{Imm}(f) = \mathfrak{A}\mathbf{C} \cap \operatorname{ran}(f).$$

Imminent Consequences

7.18 Coproduct Decomposition   Let $f \in H(A, B)$ and $g \in H(C, D)$. Then $g \in \operatorname{Imm}(f)$ implies that

$$(29) \qquad\qquad g \in H(C,D) \cap \operatorname{ran}(f).$$

But note that the existence of $g \in H(C,D) \cap \operatorname{ran}(f)$ only says

$$(30) \qquad\qquad H(C,D) \cap B \neq \varnothing$$

and does not imply $H(C,D) \subset B$, whence not $B = H(C,D)$, and therefore does not entail the conclusion $f \in H(A, H(C,D))$. This is because $f \vdash g$ only means $f(a) = g \in H(C,D)$ for an $a \in A$. It may very well happen that for another $a' \in A$, $f(a') \notin H(C,D)$; indeed, $f(a') \in \operatorname{ran}(f) \subset B$ may not even be a mapping in $\mathfrak{A}\mathbf{C}$, in which case $f(a') \notin \operatorname{Imm}(f)$, i.e. $f(a') \in \operatorname{ran}(f) \sim \mathfrak{A}\mathbf{C}$.

Since the collection of mappings $\mathfrak{A}\mathbf{C}$ is partitioned into mutually disjoint hom-sets,

$$(31) \qquad\qquad \mathfrak{A}\mathbf{C} = \bigcup_{Y,Z \in \mathbf{OC}} H(Y,Z)$$

(*cf.* Section 6.9), the set $\operatorname{Imm}(f) = \mathfrak{A}\mathbf{C} \cap \operatorname{ran}(f)$ inherits this partition, accordingly, into

$$(32) \qquad\qquad \operatorname{Imm}(f) = \mathfrak{A}\mathbf{C} \cap \operatorname{ran}(f) = \bigcup_{Y,Z \in \mathbf{OC}} \big(H(Y,Z) \cap \operatorname{ran}(f)\big).$$

This partition of $\mathrm{Imm}(f)$, in turn, may be inverse-imaged by f^{-1} into a partition of $A = \mathrm{dom}(f)$ as follows. For each ordered pair of **C**-objects (Y, Z), consider the subset X of A defined in

(33) $$X = f^{-1}\big(H(Y, Z)\big) \subset A.$$

(Note that if $g \in \mathrm{Imm}(f)$ with $f(a) = g \in H(C, D)$, then $a \in f^{-1}\big(H(C, D)\big)$.) Let the family of subsets X of the form (33) (i.e. inverse images of hom-sets) be indexed by a set I :

(34) $$\{X_i\}_{i \in I} ;$$

then

(35) $$\{f(X_i)\}_{i \in I}$$

is an indexed partition of $\mathrm{Imm}(f)$ (Section 0.19). Further, let

(36) $$A' = f^{-1}\big(\mathrm{ran}(f) \sim \mathfrak{R}\mathbf{C}\big) \subset A.$$

(It is important to remember that $f \in \mathfrak{R}\mathbf{C}$ is a standard single-valued mapping, whence

(37) $$f^{-1}\big(B \sim \mathrm{ran}(f)\big) = \varnothing$$

and

(38) $$f^{-1}(B) = f^{-1}\big(\mathrm{ran}(f)\big) = \mathrm{dom}(f) = A .)$$

Then $A = \mathrm{dom}(f)$ is the union of mutually disjoint subsets, i.e. the **Set**-*coproduct* (*cf. ML*: A.36)

(39) $$A = A' \amalg \coprod_{i \in I} X_i .$$

The mapping $f \in H(A, B)$ may then be defined 'piecewise' on the partition of its domain A, i.e. fractionated synthetically into the coproduct mapping

(40)
$$f = f|_{A'} \amalg \coprod_{i \in I} f|_{X_i} .$$

This synthesis of f from $f = \{f|_{A'}\} \cup \{f|_{X_i}\}_{i \in I}$ loses no information, since there are no further algebraic impositions on the definitions of the analytic constituent pieces. (For a detailed discussion on analysis versus synthesis, the reader is referred to Chapter 7 of *ML*: Sections 7.36 to 7.42 on 'Analytic Models and Synthetic Models' and Sections 7.43 to 7.49 on 'The Amphibology of Analysis and Synthesis'.) For an $x \in X \subset A$ with $f(x) \in H(Y,Z)$ (whence $f(x) \in \mathrm{Imm}(f)$), the *restriction* $f|_X$ of the mapping f to $X \subset A$ is of the functional entailment (i.e. repair) form

(41)
$$f|_X \in H(X, H(Y,Z)).$$

The restriction $f|_{A'}$, on the other hand, is of the material entailment (i.e. metabolism) form

(42)
$$f|_{A'} \in H(Y,Z).$$

Thus, the coproduct (40) may be considered an '(M,R)-partition' of a mapping f into its metabolism component $f|_{A'}$ and its repair component $\coprod_{i \in I} f|_{X_i}$ (with the view that 'replication' is 'repair of repair'). A mapping f for which $A' = A$ is purely metabolic; if $A' = \varnothing$, then $\mathrm{ran}(f) \subset \mathfrak{RC}$ and such a mapping f is purely repair.

One may note that although the metabolism component $f|_{A'}$ has no further imminence, that $\mathrm{ran}(f|_{A'}) \cap \mathfrak{RC} = \varnothing$, it may well be a repair component in another category, i.e. within a different set of mappings a metabolism map may have the functional entailment form and be a repair map. (See *ML*: 11.11 for (M,R)-extensions and Sections 7.5 and 7.6 for the discussion on the cyclic permutational symmetry among the (M,R)-maps.)

One may further note that the repair component $\coprod_{i \in I} f|_{X_i}$ of f entails the coproduct decomposition

(43)
$$\mathrm{Imm}(f) = \mathrm{Imm}\left(\coprod_{i \in I} f|_{X_i} \right) = \bigcup_{i \in I} \mathrm{Imm}(f|_{X_i}),$$

with each

$$(44) \qquad \mathrm{Imm}\big(f\big|_{X_i}\big) = f(X_i) \subset H(Y,Z) \subset \mathfrak{RC}$$

for some $Y, Z \in \mathfrak{OC}$. (In general, however, 'repair component maps' $f\big|_{X_i}$ may not necessarily themselves be in the collection $\mathfrak{RC}$; they are only 'fragments' of an $f \in \mathfrak{RC}$.)

So, the conclusion of this coproduct decomposition exercise is that for each mapping $f \in \mathfrak{RC}$, while it may not itself be already in the functional entailment form $f \in H\big(A, H(C,D)\big)$, there is a natural way to analyze its functional entailment parts. In particular, the repair component of each mapping $f \in \mathfrak{RC}$ may be synthesized as a coproduct of 'near-$\mathfrak{RC}$' maps $f\big|_{X_i}$ of the functional entailment form

$$(45) \qquad f\big|_{X_i} \in H\big(X_i, H(Y,Z)\big).$$

7.19 The Inverse Imm^{-1} For $f, g \in \mathfrak{RC}$, by definition of the inverse set-valued mapping (*cf.* Definition 2.13 and Lemma 2.14) $\mathrm{Imm}^{-1} : \mathfrak{RC} \multimap \mathfrak{RC}$, one has

$$(46) \qquad g \in \mathrm{Imm}^{-1}(f) \quad \text{iff} \quad f \in \mathrm{Imm}(g) \quad \text{iff} \quad g \vdash f,$$

i.e. iff g functionally entails f. Then, in particular, $\mathrm{Imm}^{-1}(f) = \varnothing$ iff $f \in \mathfrak{RC}$ is functionally entailed by no mapping in $\mathfrak{RC}$. This means any model in the category $\mathbf{C}$ (i.e. any network of mappings in $\mathfrak{RC}$) that includes an f with $\mathrm{Imm}^{-1}(f) = \varnothing$ cannot be closed to efficient causation, i.e. such a system cannot be clef. Stated otherwise, for a system N in the category $\mathbf{C}$ to be a clef, it is necessary that for every mapping $f \in \kappa(N)$, $\mathrm{Imm}^{-1}(f) \neq \varnothing$. *A fortiori,* therefore, in a clef system no mapping can be isolated under the imminence mapping Imm on $\mathfrak{RC}$ (*cf.* Definition 4.2).

7.20 The Symmetric Closure $\mathrm{Imm}^{\pm}$ The symmetric closure $\mathrm{Imm}^{\pm}$ of Imm is the set-valued mapping $\mathrm{Imm}^{\pm} = \mathrm{Imm} \cup \mathrm{Imm}^{-1} : \mathfrak{RC} \multimap \mathfrak{RC}$ defined, for $f \in \mathfrak{RC}$, by the union

$$(47) \qquad \mathrm{Imm}^{\pm}(f) = \mathrm{Imm}(f) \cup \mathrm{Imm}^{-1}(f)$$

(*cf.* Lemma 3.27 and Section 4.1). Evidently, for $f, g \in \mathfrak{RC}$,

(48) $\qquad g \in \mathrm{Imm}^{\pm}(f) \quad$ iff $\quad g \in \mathrm{Imm}(f)$ or $f \in \mathrm{Imm}(g)$,

equivalently, iff

(49) $\qquad\qquad f \vdash g$ or $g \vdash f,$

i.e. iff one of f and g functionally entails the other. As observed in Section 4.3, $g \in \mathrm{Imm}^{\pm}(f)$ iff the two mappings $f, g \in \mathfrak{H}\mathbf{C}$ are not independent (under the imminence mapping Imm on $\mathfrak{H}\mathbf{C}$). In other words, if two mappings $f, g \in \mathfrak{H}\mathbf{C}$ are independent under Imm, then neither can functionally entail the other.

Therefore, one may make the

7.21 Definition A *functionally independent set* of mappings in a category $\mathbf{C}$ is an independent set under the imminence mapping $\mathrm{Imm}_{\mathbf{C}}$ on $\mathfrak{H}\mathbf{C}$.

Whence, one has the important

7.22 Theorem *Within a functionally independent set, no clef system can exist.*

7.23 Corollary *Within a functionally independent set, no (M,R)-system can exist.*

Since (M,R)-systems define organisms, this corollary gives a formal condition under which no life forms can arise, a prescription of a sterile world. This is the reason why the concept of independent sets is important in relational biology, and wherefore, in Chapter 4, I have considered in detail some conditions under which one may account for the number and cardinality of independent sets.

As an illustrative example, Theorem 4.13, when specialized to the imminence mapping Imm on $\mathfrak{H}\mathbf{C}$, becomes

7.24 Theorem *Let $\mathfrak{H}\mathbf{C}$ be a finite or countably infinite set. If there exists a nonnegative integer n such that, for each $f \in \mathfrak{H}\mathbf{C}$, $\left|\mathrm{Imm}_{\mathbf{C}}(f)\right| \le n$, then $\mathfrak{H}\mathbf{C}$ may be partitioned into $2n+1$ independent sets.*

Thus, one sees that under certain finite-bound conditions on $\mathrm{Imm}_{\mathbf{C}}$, the collection $\mathfrak{H}\mathbf{C}$ of $\mathbf{C}$-morphisms may be partitioned into the union of a finite number of functionally independent sets, whence (as noted in 4.18) into the union of a finite number of maximal functionally independent sets.

8
Imminence of Life

... (les) questions les plus importantes de la vie ... ne sont en
effet, pour la plupart, que des problèmes de probabilité.

[The most important questions of life are in effect, for the most
part, nothing but probability problems.]

> — Pierre-Simon Laplace (1820)
> *Théorie Analytique des Probabilitiés*
> (*Oeuvres complètes*, t. 7) Introduction

Genesis of Functional Entailment

8.1 Fiat Intuitively, in a category **C**, if there is only a small finite number of
large (perhaps even of infinite cardinality) functionally independent sets of
mappings, then the probability of a collection of mappings from ℛ**C** to
functionally entails one another is small. Contrariwise, if the number of maximal
functionally independent sets of mappings is large and their cardinalities are small,
then a collection of mappings from ℛ**C** is less likely to all be from the same
functionally independent set, whence mutual functional entailments of the
mappings are feasible.

An application of the *Theoria simiarum infinitarum* may be used to infer
inevitability from probability. This propensity for the emergence of functional
entailment may, indeed, be made more rigorous into probabilistic theorems. The
Second Borel–Cantelli Lemma states that if the events in an infinite sequence are
independent and the sum of their probabilities diverges to infinity, then the
probability that infinitely many of them occur is 1. This lemma is the basis of all
theorems of the strong type (i.e. on 'almost sure convergence') in probability
theory. For our purpose of 'origin of functional entailment', the pertinent
corollary of the lemma is that if the occurrence of an event has constant nonzero
probability (however small), then the event will almost surely occur in an infinite
sequence of independent repetitions.

A.H. Louie, *The Reflection of Life*, IFSR International Series on Systems Science
and Engineering 29, DOI 10.1007/978-1-4614-6928-5_8,
© Springer Science+Business Media New York 2013

The important conclusion here is that in a suitably equipped category **C** (in the sense of an appropriate selection of mappings to belong to the collection $\mathcal{H}C$ of **C**-morphisms), any sufficiently large finite family of morphisms must inevitably contain a mutual connection that is functional entailment. Recall that the essence of an (M,R)-system is the

(1)   'repair $\vdash$ metabolism'

functional entailment. *A becoming imminence mapping* Imm_C *on* $\mathcal{H}C$ *coming into being* is thus the *dies unus* of genesis.

8.2 Towards Complexity   The genesis of functional entailment from a random collection of mappings–processes is the first step of emergence in entailment networks. The next step in the relational development of entailment networks is when a hierarchical chain folds into a cycle, in the creation of a *complex system* (*cf. ML*: Chapter 9). The study of when this happens is a topic in graph theory (*cf. ML*: Chapter 6) through relational digraphs (*cf.* Section 3.34 *et seq.*), in particular when specialized to the digraph corresponding to the imminence mapping $\text{Imm} : \mathcal{H}C \multimap \mathcal{H}C$.

Recall that cycles may also be explored through the *adjacency matrix* $[G]$ (Section 5.2) of an entailment network G : the diagonal entry $\left(G^k\right)_{ii}$ of the power $\left[G^k\right]$ indicates the number of cycles of length k in the graph G from vertex i to itself.

Further Graph Theory

In graph theory, one is often interested in determining whether a given graph contains a subgraph with a prescribed property. For our present endeavour, we are most concerned with conditions under which a graph contains a cycle. There are many theorems and conjectures in this subject area; a good reference is Bondy and Murty [2008]. I list a few sample results to show the cornucopia.

8.3 Lemma   *A graph with at least as many edges as vertices contains a cycle. Every n-vertex graph with n+1 edges contains at least two (possibly overlapping) cycles.*

8.4 Lemma   *Let G be a graph in which all vertices have degree at least two. Then G contains a cycle.*

8.5 Lemma   *Let the average degree of a connected graph G be greater than two. Then G has at least two cycles.*

8.6 Theorem   *Let G be a graph in which every vertex has even degree; then there exists a collection of cycles of G so that every edge appears in exactly one of these cycles.*

8.7 Theorem   *Let the simple graph G have minimum degree δ (i.e. all vertices have degree $\geq \delta$). If $\delta \geq 2$, then G contains a cycle of length (i.e. with number of edges) at least $\delta + 1$.*

8.8 Theorem   *Let the simple digraph G have minimum degree δ (i.e. all vertices have indegree $\geq \delta$ and outdegree $\geq \delta$). If $\delta > 0$, then G contains a directed cycle of length at least $\delta + 1$.*

Random Graphs

An understanding of the expected behaviour of graphs, of 'how graphs behave on average', is of immense value in demonstrating the existence of graphs with particular properties. The 'probabilistic method' in graph theory, called the theory of *random graphs*, is also a remarkably effective tool for characterizing graphs in general.   Paul Erdős was the master of combinatorial mathematics and the probabilistic method (we have already encountered some of his expert counting in Chapter 4).   It is therefore not surprising that he had also started (along with Alfréd Rényi) the study of random graphs [Erdős & Rényi 1959, 1960].

Let me ease into the subject with a quick review of a few standard concepts in stochastics, mostly for the notation.

8.9 Definition   A (*finite*) *probability space* (Ω, P) consists of a finite set Ω, called the *sample space*, and a *probability function* $P : \Omega \to [0,1]$ satisfying $\sum_{\omega \in \Omega} P(\omega) = 1$.

8.10 Event   In a probability space (Ω, P), any subset $A \subset \Omega$ is referred to as an *event*, and the *probability* of the event A is defined by $P(A) = \sum_{\omega \in A} P(\omega)$. For each event A, $0 \leq P(A) \leq 1$; $P(\varnothing) = 0$ and $P(\Omega) = 1$.

8.11 Random Variable   A *random variable* X on a probability space (Ω, P) is any real-valued mapping $X : \Omega \to \mathbb{R}$.   An event A in a probability space (Ω, P) has an associated *indicator random variable* X_A, defined by

$$(2) \qquad X_A(\omega) = \begin{cases} 1 & \text{if } \omega \in A \\ 0 & \text{otherwise} \end{cases}.$$

8.12 Expectation   The average value, or mean, of a random variable X is called its *expectation* (also *expected value*), denoted by $E(X)$:

$$(3) \qquad E(X) = \sum_{\omega \in \Omega} X(\omega)P(\omega).$$

The mapping

$$(4) \qquad E \in H\big(H(\Omega,\mathbb{R}),\mathbb{R}\big)$$

is *linear*: for random variables X and Y and real numbers a and b,

$$(5) \qquad E(aX+bY) = aE(X)+bE(Y).$$

For an indicator random variable X_A,

$$(6) \qquad E(X_A) = P(X_A = 1).$$

8.13 Markov's Inequality   *Let X be a nonnegative random variable and t a positive real number. Then*

$$(7) \qquad P(X \geq t) \leq \frac{E(X)}{t}.$$

8.14 Corollary   *For $n = 1,2,3,\ldots$, let X_n be a nonnegative integer-valued random variable on a probability space (Ω_n, P_n). If $E(X_n) \to 0$ as $n \to \infty$, then $P(X_n = 0) \to 1$ as $n \to \infty$.*

8.15 Variance   It is often useful to know how concentrated a random variable's distribution is about its expectation. This is measured by the *variance* X, defined by

$$(8) \qquad V(X) = E\big((X - E(X))^2\big).$$

The smaller the variance, the more concentrated the random variable is about its expectation.

Now let us see how the probabilistic method enhances graph theory.

8.16 The Probability Space $G_{n,p}$ Consider the sample space G_n of all simple graphs with n vertices. There are $\binom{n}{2} = \frac{1}{2}n(n-1)$ vertex pairs among the n, and each pair is either connected by an edge or not; thus,

(9) $$|G_n| = 2^{\frac{1}{2}n(n-1)}.$$

In a finite probability space (G_n, P), the result of selecting an element $G \in G_n$ according to the probability function P is called a *random graph*.

Of particular interest is the general probability function P defined as follows. Fix a real number $p \in [0,1]$ and define the probability of choosing each edge as p (with the choices independent of one another), whence the probability of not choosing an edge is $1 - p$. The resulting probability function

(10) $$P : G_n \to [0,1]$$

defined for $G \in G_n$ where G has m edges ($0 \le m \le \frac{1}{2}n(n-1)$) is thus

(11) $$P(G) = p^m (1-p)^{\frac{1}{2}n(n-1)-m}.$$

The probability space (G_n, P) with this special p-dependent probability function P is denoted $G_{n,p}$.

8.17 Properties   A *property* Π of graphs with n vertices may be considered the subset Π_n of G_n (i.e. the event) defined by this property (*cf.* Axiom of Specification; *ML*: 0.19):

(12) $$\Pi_n = \{G \in G_n : \Pi(G)\}.$$

A *monotone (increasing) property* of graphs is one which is preserved when edges are added. This is to say, for a monotone property Π, if $G_1, G_2 \in G_n$ and G_1 is a subgraph of G_2, then $G_1 \in \Pi_n$ implies $G_2 \in \Pi_n$. Examples of monotone increasing properties are 'a fixed graph G_0 appears as a subgraph in G', 'there exists a large component of a fixed size in G', and 'G is connected'.

Consider a sequence of probability spaces $\{G_{n,p}\}_{n\in\mathbb{N}}$ with the probability p dependent on n; i.e. probability p is a mapping $p:\mathbb{N}\to[0,1]$. One says that *almost every* (*a.e.*) graph in $G_{n,p}$ has the monotone property Π if

$$(13) \qquad\qquad\qquad P(\Pi_n) \to 1 \quad \text{as} \quad n\to\infty.$$

Note that, in the limit $G_{\mathbb{N},p} = \bigcup_{n\in\mathbb{N}} G_{n,p}$, the definition of 'almost every' coincides with that in measure theory: a. e. $G \in G_{\mathbb{N},p}$ has the property Π if $P(G\in\Pi)=1$.

8.18 Triangles   As a simple example, let X be the random variable that counts the number of triangles (i.e. cycles of length 3) in a graph $G \in G_{n,p}$. Let S be a 3-vertex subgraph of G (with all the G-edges, if they exist, connecting these three vertices). Let A_S be the event 'S is a triangle', and X_{A_S} be its associated indicator random variable. One sees that

$$(14) \qquad\qquad E\left(X_{A_S}\right) = P\left(X_{A_S}=1\right) = P\left(A_S\right) = p^3.$$

X may be expressed as the sum

$$(15) \qquad\qquad\qquad X = \sum X_{A_S}$$

over all 3-vertex subgraphs S of G. There are $\binom{n}{3}$ such subgraphs, whence

$$(16) \qquad\qquad E(X) = \sum E\left(X_{A_S}\right) = \binom{n}{3}p^3 < (pn)^3.$$

Thus, if $pn\to 0$ as $n\to\infty$, then $E(X)\to 0$ and, by Markov's Inequality 8.14, $P(X=0)\to 1$; in other words, if $pn\to 0$ as $n\to\infty$, then almost every graph $G \in G_{n,p}$ is triangle-free.

The following inequality is a corollary of Markov's Inequality, and the two inequalities play, in some sense, complementary roles to each other.

8.19 Chebyshev's Inequality   *Let X be a random variable and t a positive real number. Then*

(17)
$$P\big(\big|X - E(X)\big| \geq t\big) \geq \frac{V(X)}{t^2}.$$

Chebyshev's Inequality bounds the divergence of a random variable from its mean; its more useful form for random graph theory is

8.20 Corollary *For* $n = 1, 2, 3, \ldots$, *let* X_n *be a random variable on a probability space* (Ω_n, P_n) . *If* $E(X_n) \neq 0$ *and* $V(X_n)\big/\big(E(X_n)\big)^2 \to 0$ *as* $n \to \infty$, *then* $P(X_n = 0) \to 0$ *as* $n \to \infty$.

8.21 More Triangles Using arguments similar to (although slightly more complicated than) those in Section 8.18, it may be shown, using Chebyshev's Inequality 8.20, that if $pn \to \infty$ as $n \to \infty$, then almost every graph $G \in \mathsf{G}_{n,p}$ will contain at least one triangle.

In summary, one sees that the behaviour of a random graph $G \in \mathsf{G}_{n,p}$, for the defining probability mapping $p \in [0,1]^{\mathbb{N}}$ of the sequence of probability spaces $\{\mathsf{G}_{n,p}\}_{n \in \mathbb{N}}$, bifurcates at the *critical point* n^{-1}: if, as $n \to \infty$, $p(n)\big/n^{-1} = pn \to 0$, then almost every graph G has no triangles, whereas if $p(n)\big/n^{-1} = pn \to \infty$, then almost every graph G has at least one triangle. One may call the mapping $\tau(n) = n^{-1}$ a *threshold function* for the property of 'containing a triangle'.

8.22 Threshold For a monotone property Π, a *threshold function* for Π is a mapping $\tau : \mathbb{N} \to \mathbb{R}$ such that, as $n \to \infty$, if $p(n)/\tau(n) \to 0$, then almost no $G \in \mathsf{G}_{n,p}$ has Π (i.e. almost every $G \in \mathsf{G}_{n,p}$ has $\neg\Pi$, the negation of Π), and if $p(n)/\tau(n) \to \infty$, then almost every $G \in \mathsf{G}_{n,p}$ has Π. Threshold functions are not unique (whence the indefinite article in its definition). For example, for any constant $c > 0$, $\tau(n) = cn^{-1}$ is also a threshold function for the property Π of 'containing a triangle'. It turns out that $\tau(n)$ is a threshold if and only if any mapping that has similar asymptotic behaviour to $\tau(n)$ is a threshold; it is in this 'equivalence in asymptotic behaviour' sense that one may sometimes speak of *the* threshold function of a property.

It is an important result that the same threshold function for triangles is also for the appearance of cycles of all lengths:

8.23 Theorem *The mapping* $\tau(n) = n^{-1}$ *is a threshold function for the appearance in a random graph in* $\{\mathsf{G}_{n,p}\}_{n \in \mathbb{N}}$ *of cycles (of any fixed length).*

Another important property of a graph is connectedness:

8.24 Theorem   *The mapping* $\tau(n) = \log n / n$ *is a threshold function for a random graph in* $\{G_{n,p}\}_{n \in \mathbb{N}}$ *to be connected.*

Spiritus Vitae

8.25 Threshold Crossing   One may intuitively consider the behaviour of random graphs $\{G_{n,p}\}_{n \in \mathbb{N}}$ a *developmental process*, with random acquisition of edges by $G \in G_{n,p}$. One of the main aims of the theory of random graphs is to determine when a given property is likely to appear. Erdős and Rényi discovered that the appearance of most monotone properties is sudden ('emergence'). Threshold functions define behavioural bifurcations ('phase transitions') at their critical points, before which the property is unlikely ('meagre') and after which it is very likely ('generic').

For our purposes, the appearance of cycles is the next important emergence in network development: this is when simple becomes complex. It is important to note that this threshold crossing is not a rote algorithmic operation of adding more of the same but a fundamental change in network topology. The threshold mapping $\tau(n)$ partitions natural systems into complementary collections that are different in kind in terms of their functional entailment networks. Beyond a threshold in the developmental process, among natural systems, simple systems are meagre and complex systems are generic (*cf. ML*: 9.6).

8.26 Incompleteness as Metaphor   Kurt Gödel's incompleteness theorems are two theorems in mathematical logic about inherent limitations of axiomatic systems capable of doing a certain amount of elementary arithmetic. The first incompleteness theorem states that no consistent formal system in which theorems can be derived by an algorithmic procedure is complete. Stated otherwise, for any system within which a certain nontrivial amount of elementary arithmetic can be carried out, there will always be statements about the natural numbers that are true but are unprovable within the system. The second incompleteness theorem may be considered a corollary of the first and shows that such a system's own consistency cannot be demonstrated within the system itself. One may note in passing that the 'certain nontrivial amount of elementary arithmetic' requisite in the formal systems are not the same in the two theorems.

The two theorems, proven by Gödel in 1931, are important both in mathematical logic and in the philosophy of mathematics. In essence, the first theorem implies that if something relatively simple like arithmetic already contains such foundational difficulties, then an all-encompassing axiomatic system can never be found that is able to prove *all* mathematical truths (and nothing but those truths). The contrapositive form of the second theorem says that if an axiomatic system can be proven to be consistent from within itself, then it is inconsistent. These interpretations show that David Hilbert's proposed program to find a complete and consistent set of axioms for all mathematics is impossible.

Analogies have often been made to the incompleteness theorems in support of arguments in scholastic pursuits beyond mathematics and logic. Consequently, there has sprung a mini-industry of publications by authors (notably indignant logicians) that have reacted negatively on such 'contaminating' extensions. To be sure, Gödel's incompleteness theorems are *about* number theory, and statements of their invocation in any other fields are not proven. Just as surely, however, the lessons to be learned from Gödel are more importantly the metaphorical ones. One such lesson (as I have written in *ML*: 7.48) of incompleteness is that "nothing works for everything": all attempts at universality or genericity in human endeavours are likely unsuccessful, indeed an *in frustra* transgression.

Let me point out another analogy. Both incompleteness theorems require the formal system to be capable of carrying out 'a certain amount of elementary arithmetic'. For the impredicativity in the conclusions to emerge, the axiomatic systems cannot be 'too simple'. The condition that is 'a certain amount of elementary arithmetic' has to do with what can be expressed and what can be proven within a system and may metaphorically be considered a measure of *complexity*. Note that an appropriate sense of the word 'complexity' must be defined (*cf. ML*: 9.1); a 'complicated' simple system is still simple. The lesson of incompleteness in this context is, therefore, when the increasing complexity of a system reaches a certain level (e.g. when the interconnection of processes in an entailment network exceeds a certain 'threshold'), impredicativity is an inevitable outcome.

8.27 Whence Clef   We now have a theory for the first two steps in the development of entailment networks: out of a collection of processes, functional entailment emerges (Section 8.1) and, among mutually interacting processes, hierarchical cycles are formed after crossing a certain threshold (Theorem 8.23).

The third step is *complexitas viventia producit*: 'complexity brings forth living beings'. This is the synthetic step when functional entailment networks achieve process closure, when mutual functional entailment attains self-sufficiency and when (M,R)-networks become (M,R)-systems (*cf. ML*: 11.27 *et seq.* and *ML*: 13.23 *et seq.*).

Let N be a natural system modelled in the category $\mathbf{C}$ and $\kappa(N) \subset \mathfrak{A}\mathbf{C}$ be its collection of efficient causes. For $f \in \kappa(N) \subset \mathfrak{A}\mathbf{C}$, the inverse image of f under the imminence mapping $\mathrm{Imm}: \mathfrak{A}\mathbf{C} \multimap \mathfrak{A}\mathbf{C}$ of the category $\mathbf{C}$, the subset $\mathrm{Imm}^{-1}(f) \subset \mathfrak{A}\mathbf{C}$, contains all the mappings in $\mathfrak{A}\mathbf{C}$ that functionally entail f (*cf. Section 7.19*). If $\mathrm{Imm}^{-1}(f) = \varnothing$, then f is not functionally entailed by any $\mathbf{C}$-morphism, whence the system N cannot be closed to efficient causation. Even when $\mathfrak{A}\mathbf{C}$ is sufficiently networked in functional entailment so that $\mathrm{Imm}^{-1}(f) \neq \varnothing$, it may very well happen that $\mathrm{Imm}^{-1}(f) \subset \mathfrak{A}\mathbf{C} \sim \kappa(N)$, whence f is functionally entailed by $\mathbf{C}$-morphisms external to N, and again, the system

N is not closed to efficient causation. One sees, therefore, for the system N to be clef, one must have, for *every* $f \in \kappa(N)$,

(18) $\mathrm{Imm}^{-1}(f) \cap \kappa(N) \neq \varnothing$.

This is a stringent requirement, but not prohibitively so. The requirement may be satisfied when the category **C** is 'suitably equipped': the equipment depends on the availability of maps in the category **C**, the mutual entailment pattern in the collection $\mathfrak{H}\mathbf{C}$ of **C**-morphisms, and the partitioning of the natural system N from its external world, its complement N^c (i.e. the partitioning of $\mathfrak{H}\mathbf{C}$ into $\kappa(N)$ and $\mathfrak{H}\mathbf{C} \sim \kappa(N)$).

8.28 Relational Genesis   The biblical account of Genesis is quite reductionistic, and the narration emphasizes the physicochemical material causes: universe, light, sky, land, sea, sun, moon, and so on. Plants (*'herbam virentem'*) appear on the third day, and animals (*'animam viventem'*) begin to appear on the fifth day.

Our relational account of genesis is formulated in terms of efficient causes, on the development of the imminence mapping $\mathrm{Imm}_{\mathbf{C}} : \mathfrak{H}\mathbf{C} \multimap \mathfrak{H}\mathbf{C}$ of a category **C**. The three emergent steps in relational networks N are as follows:

i.   Functional entailment: $\mathrm{Imm}_{\mathbf{C}}(f) \neq \varnothing$

ii. Hierarchical cycle in the digraph $\mathrm{Imm}_{\mathbf{C}}$: crossing the threshold $\tau(n)$

iii. Clef: for every $f \in \kappa(N)$, $\mathrm{Imm}_{\mathbf{C}}^{-1}(f) \cap \kappa(N) \neq \varnothing$

The apparent improbability of such a set of events becomes probable (and inevitable) under strong laws of almost sure convergence in probability theory. As I have explained in Section 8.1, all that is required is for the sum of the nonzero probabilities of an infinite sequence of independent events to diverge; then the events will almost surely occur.

'Almost surety' is, however, not absolute certainty. All three emergent steps of relational development are stochastic entailments: they are only nearly inevitable. But note that genesis is an event that needs to happen only once (although there is nothing that prohibits its recurrence). Contrariwise, the continuous process of evolution, the adaptive fit of species to their various environments, is cumulative and ongoing.

Now, from the standpoint of relational biology, based on the emerging complexity of the imminence mapping $\mathrm{Imm}_{\mathbf{C}} : \mathfrak{H}\mathbf{C} \multimap \mathfrak{H}\mathbf{C}$ of the category **C**, I submit the following statement on the origin of life:

8.29 Postulate of Biopoiesis   In a suitably equipped category, any sufficiently large finite family of morphisms must contain an (M,R)-system.

The Postulate of Biopoiesis is the 'becoming' counterpart of our 'being' postulate:

8.30 Postulate of Life (*ML*: 11.28)   A natural system is an organism if and only if it realizes an (M,R)-system.

These two statements 8.29 and 8.30, in other words, are, respectively, the postulates of ontogeny and epistemology of life in relational biology.

8.31 Relational Biology   The subject of the Rashevsky–Rosen school is mathematical biology. Relational biology is the operational description of our endeavour, the characteristic name of our approach. Relational biology can no more be done without the mathematics than without the biology, mathematics and biology being the two complementary halves of our subject. That relational organizations and not physicochemical structures define life is, of course, the motto. But, one must understand that the 'relational' in 'relational biology' is not just used in its common-usage sense of 'having an effect of a connection' (sometimes even misinterpreted as 'relative'). 'Relational' is more importantly used in its mathematical sense (Definition 1.3), that 'a (mathematical) relation exists'.

In *Topology and Life* [Rashevsky 1954], where relational biology began, Nicolas Rashevsky entwined the two complementary halves of our subject thus:

> The appropriate representation of the relations between the different biological functions of an organism appears to be a one-dimensional complex, or graph, which represents the "organization chart" of the organism.

I now declare that this "organization chart" is the imminence mapping, the relation $\mathrm{Imm}_c \subset \mathfrak{R}C \times \mathfrak{R}C$.

9
Imminence in Models

Intention, *n.* The mind's sense of the prevalence of one set of influences over another set; an effect whose cause is the imminence, immediate or remote, of the performance of an involuntary act.

> — Ambrose Bierce (1911)
> *The Devil's Dictionary*
> (*The Collected Works of Ambrose Bierce*, vol. VII)

Imminence Mapping of a System

9.1 Imminence Mapping of a Model   The 'global' theory of the imminence mapping $\mathrm{Imm}_{\mathbf{C}} : \mathfrak{A}\mathbf{C} \multimap \mathfrak{A}\mathbf{C}$ of the category $\mathbf{C}$ may be restricted 'regionally' to a model of a natural system N in the category $\mathbf{C}$ (alternatively, a system N in the category $\mathbf{C}$). Instead of the whole collection $\mathfrak{A}\mathbf{C}$ of $\mathbf{C}$-morphisms, consider a formal system N that is a network of mappings in $\mathfrak{A}\mathbf{C}$ (e.g. an (M,R)-network), whence the collection $\kappa(N)$ of all efficient causes in N is a subset of $\mathfrak{A}\mathbf{C}$, viz., $\kappa(N) \subset \mathfrak{A}\mathbf{C}$ (*cf.* Sections 7.10 and 7.14). The *imminence mapping of the system N in the category* $\mathbf{C}$ (also the *imminence mapping on* $\kappa(N)$) is the set-valued mapping

(1)
$$\mathrm{Imm}_N : \kappa(N) \multimap \kappa(N)$$

defined, for each mapping $f \in \kappa(N)$, by

(2)
$$\mathrm{Imm}_N(f) = \kappa(N) \cap \mathrm{ran}(f).$$

A.H. Louie, *The Reflection of Life*, IFSR International Series on Systems Science and Engineering 29, DOI 10.1007/978-1-4614-6928-5_9, © Springer Science+Business Media New York 2013

Imm_N is a relation on $\kappa(N)$ and a member of the monoid $\mathbf{Rel}\big(\kappa(N),\kappa(N)\big)$ (*cf.* Section 3.21). The set $\text{Imm}_N(f)$ is the collection of all efficient causes of N that lie in the range of f, i.e. all the f-entailed entities in $\kappa(N)$. The imminence mapping Imm_N on $\kappa(N)$ is the functional entailment pattern of the model of the natural system N and may therefore be taken as the very definition of the entailment network that is the natural system N.

Since $\kappa(N)\subset\mathfrak{AC}$, for each mapping $f\in\kappa(N)$,

$$(3) \qquad \text{Imm}_N(f)\subset\text{Imm}_{\mathbf{C}}(f) \quad\text{and}\quad \text{Imm}_N^{-1}(f)\subset\text{Imm}_{\mathbf{C}}^{-1}(f).$$

Indeed, $\text{Imm}_N:\kappa(N)\multimap\kappa(N)$ and $\text{Imm}_N^{-1}:\kappa(N)\multimap\kappa(N)$ may be considered the *restriction* of, respectively, $\text{Imm}_{\mathbf{C}}:\mathfrak{AC}\multimap\mathfrak{AC}$ and $\text{Imm}_{\mathbf{C}}^{-1}:\mathfrak{AC}\multimap\mathfrak{AC}$ to the domain and codomain $\kappa(N)\subset\mathfrak{AC}$; i.e. Imm_N, $\text{Imm}_N^{-1}\in\mathbf{Rel}\big(\kappa(N),\kappa(N)\big)$ and $\text{Imm}_{\mathbf{C}}$, $\text{Imm}_{\mathbf{C}}^{-1}\in\mathbf{Rel}\big(\mathfrak{AC},\mathfrak{AC}\big)$ with $\text{Imm}_N\subset\text{Imm}_{\mathbf{C}}$ and $\text{Imm}_N^{-1}\subset\text{Imm}_{\mathbf{C}}^{-1}$ on $\kappa(N)\subset\mathfrak{AC}$. In Section 8.27, we saw that the requirement for the system N to be closed to efficient causation is, for every $f\in\kappa(N)$, $\text{Imm}_{\mathbf{C}}^{-1}(f)\cap\kappa(N)\neq\varnothing$. Thus, it follows immediately that one has the

9.2 Theorem   *A system N is clef if and only if, for every $f\in\kappa(N)$,* $\text{Imm}_N^{-1}(f)\neq\varnothing$.

9.3 Functional Closure   The necessary and sufficient condition for a nonempty system N to be a clef system, that for every $f\in\kappa(N)$, $\text{Imm}_{\mathbf{C}}^{-1}(f)\cap\kappa(N)=\text{Imm}_N^{-1}(f)\neq\varnothing$, means that there exists an efficient cause g_f (that depends on f) in N such that $g_f\in\text{Imm}_N^{-1}(f)$. Now consider the family of nonempty sets $\big\{\text{Imm}_N^{-1}(f)\big\}_{f\in\kappa(N)}$ indexed by the nonempty set $\kappa(N)$. One may invoke the Axiom of Choice (Axioms 0.20 and 1.2) and conclude that there exists an indexed family $\big\{g_f\big\}_{f\in\kappa(N)}$ of processes in N such that for each $f\in\kappa(N)$, $g_f\in\text{Imm}_N^{-1}(f)$. (The invocation is, in fact, not mandatory, since the nonempty set $\kappa(N)$ of processes in N is finite; *cf.* ML: 7.20: Axiom of Finitude.)

Define the choice mapping $\varepsilon:\big\{\text{Imm}_N^{-1}(f):f\in\kappa(N)\big\}\to\kappa(N)$ by $\varepsilon\big(\text{Imm}_N^{-1}(f)\big)=g_f$, and define the mapping $\lambda:\kappa(N)\to\kappa(N)$, for $f\in\kappa(N)$, by the sequential composite of the set-valued mapping $\text{Imm}_N^{-1}:\kappa(N)\multimap\kappa(N)$

(interpreted as the single-valued mapping $\text{Imm}_N^{-1} : \kappa(N) \to P(\kappa(N))$; *cf.* Definition B in Section 2.1) and the choice mapping ε :

(4) $$\lambda(f) = (\varepsilon \circ \text{Imm}_N^{-1})(f) = \varepsilon(\text{Imm}_N^{-1}(f)) = g_f .$$

The definition (4) of the mapping $\lambda : \kappa(N) \to \kappa(N)$ may appear convoluted. It may, however, be simply explained as a mapping that chooses, for each efficient cause f of the clef system N, a functional entailer, i.e. an efficient cause g_f within the system N that entails f :

(5) $$\lambda : f \mapsto g_f \text{ and } g_f \vdash f .$$

Through the power set functor (*cf.* Section 1.18), $\lambda : \kappa(N) \to \kappa(N)$ defines the mapping $P\lambda : P(\kappa(N)) \to P(\kappa(N))$ of powers sets: for each $A \subset \kappa(N)$,

(6) $$P\lambda(A) = \lambda(A) = \{g_f : f \in A\} .$$

Thus, $P\lambda$ is a unary operation on $P(\kappa(N))$ (*cf.* Section 3.23). Let $\mathfrak{S} \subset P(\kappa(N))$ be the collection of all singleton subsets of $\kappa(N)$; i.e.

(7) $$\mathfrak{S} = \{\{f\} : f \in \kappa(N)\} .$$

Then

(8) $$P\lambda(\mathfrak{S}) = \{P\lambda(\{f\}) : f \in \kappa(N)\} = \{\{g_f\} : f \in \kappa(N)\} .$$

Now, each $\{g_f\}$ is a singleton subset of $\kappa(N)$, i.e. $\{g_f\} \in \mathfrak{S}$. $P\lambda(\mathfrak{S})$ is a collection of such; therefore,

(9) $$P\lambda(\mathfrak{S}) \subset \mathfrak{S} .$$

This says $\mathfrak{S}$ *is closed under the unary operation* $P\lambda$ *on* $P(\kappa(N))$ (*cf.* Section 3.23).

In summary, we have linked the 'closure' in 'closure in efficient causation' for a clef system to the general 'operation closure' of Section 3.23, whence explained the word's usage:

9.4 Theorem   *A system N is closed to efficient causation if and only if the collection of all singleton subsets of its efficient causes is closed under the unary operation that chooses a functional entailer for each efficient cause.*

Imminence in (M,R)-Networks

By Definition 2.4, $\mathrm{Imm}_N(f) \neq \varnothing$ iff $f \in \mathrm{cor}\left(\mathrm{Imm}_N\right)$. A mapping $f \in \mathrm{cor}\left(\mathrm{Imm}_N\right)$ functionally entails other mappings in $\kappa(N)$; such an f is a repair map. A subset $A \subset \mathrm{cor}\left(\mathrm{Imm}_N\right)$ of mappings in $\kappa(N)$ is therefore a collection of repair maps, and $\mathrm{Imm}_N(A) \subset \kappa(N)$ is the imminence of A, the collection of mappings in $\kappa(N)$ that A functionally entails. The set $\mathrm{Imm}_N(A)$ contains, in particular, all the metabolism maps that correspond to (i.e. that are functionally entailed by) mappings in A. Inversely, for a subset $B \subset \kappa(N)$ of mappings of the system N, the set $\mathrm{Imm}_N^{-1}(B)$ contains all the mappings in $\kappa(N)$ that functionally entail B. When specialized to the imminence mapping on $\kappa(N)$, Lemma 2.21 becomes

9.5 Theorem $A \subset \mathrm{Imm}_N^{-1}\left(\mathrm{Imm}_N(A)\right)$ *iff* $A \subset \mathrm{cor}\left(\mathrm{Imm}_N\right)$.

Recall (Theorem 1.21.i) that for a (standard single-valued) mapping $f : X \to Y$, for all subsets $A \subset X$, one has $A \subset f^{-1}(f(A))$. But for set-valued mappings, further containment restrictions have to be satisfied before the 'invert-and-recover' operation succeeds. Theorem 9.5 is thus an important characterization of repair maps, as those mappings that may be *recovered from their imminence*, i.e. entailed from their corresponding metabolism maps via the inverse imminence mapping. The containment condition $A \subset \mathrm{Imm}_N^{-1}\left(\mathrm{Imm}_N(A)\right)$ serves to distinguish collections A of repair maps from those that contain metabolism maps.

9.6 Specificity   Let $f, g, h \in \kappa(N)$ with $f \vdash h$ and $g \vdash h$, i.e. $h \in R_N(f)$ and $h \in R_N(g)$. Then $\{f, g\} \subset R_N^{-1}(h)$, whence $\{f\} \neq \mathrm{Imm}_N^{-1}\left(\mathrm{Imm}_N\{f\}\right)$ (and $\{g\} \neq \mathrm{Imm}_N^{-1}\left(\mathrm{Imm}_N\{g\}\right)$). This simple illustration shows that if a mapping is functionally entailed by two different mappings, then the 'invert-and-recover' operation is not specific (i.e. $\mathrm{Imm}_N^{-1} \circ \mathrm{Imm}_N$ is not the identity mapping $1_{\kappa(N)}$). If Imm_N is injective (*cf.* Definition 2.6), however, then its defining property $f \neq g \;\Rightarrow\; \mathrm{Imm}_N(f) \cap \mathrm{Imm}_N(g) = \varnothing$ means such a two-to-one functional entailment cannot happen. A natural system N with an injective imminence mapping Imm_N thus has inherent 'uniqueness' properties, which may, when N is an (M,R)-network, be interpreted in various biological contexts such as enzyme specificity and one-gene-one-enzyme determinacy.

Recall (Definition 2.22) that for $F : X \multimap Y$, a subset $A \subset X$ for which $A = F^{-1}(F(A))$ is special and is given the special name of a stable subset (of X under F). In the context of the imminence mapping Imm_N of the system N, one may give the following:

9.7 Definition A subset $A \subset \kappa(N)$ for which $A = \mathrm{Imm}_N^{-1}\left(\mathrm{Imm}_N(A)\right)$ is called a *functionally stable subset* (of the system N).

It follows from Theorem 9.5 that a functionally stable subset must be a subset of the corange $\mathrm{cor}\left(\mathrm{Imm}_N\right)$. Thus, Theorem 2.23 specializes into

9.8 Theorem *The functionally stable subsets form a complemented lattice (a complemented sublattice of the power set lattice* $P\left(\mathrm{cor}\left(\mathrm{Imm}_N\right)\right)$ *).*

In terms of (M,R)-networks, functionally stable subsets contain repair components for which the 'repair $\vdash$ metabolism' entailment is 'robust' (indeed, 'functionally stable', hence the name), in the sense that a specific functional entailment $f \vdash h$ is not readily perturbed ('hijacked') into another $g \vdash h$. I shall have more to say about functional stability in the context of pathophysiology in Chapter 12.

Iterated Imminence

9.9 Sequential Imminence For $f \in \kappa(N)$, the squared imminence is the sequential composite $\mathrm{Imm}_N^2 = \mathrm{Imm}_N \circ \mathrm{Imm}_N : \kappa(N) \multimap \kappa(N)$ defined by

$$(10) \quad \mathrm{Imm}_N^2(f) = \left(\mathrm{Imm}_N \circ \mathrm{Imm}_N\right)(f) = \bigcup_{g \in \mathrm{Imm}_N(f)} \mathrm{Imm}_N(g) \subset \kappa(N)$$

(*cf.* Definition 3.4.i and Section 3.21).

A mapping $h \in \mathrm{Imm}_N^2(f)$ is reachable from the mapping f after travelling on two connected pairs of solid-headed and hollow-headed arrows in the digraph representation of Imm_N (*cf.* Section 3.36). The iterated imminence $h \in \mathrm{Imm}_N^2(f)$ entails the existence of an intermediary mapping $g \in \mathrm{Imm}_N(f)$ in the imminence of f such that

(11) $f \vdash g \vdash h.$

Symbolically, this situation may be summarized

(12) $h \in \mathrm{Imm}_N^2(f) \;\vdash\; \left(\exists g : f \vdash g \vdash h\right).$

Recall (Section 7.12) that the entailment of the existence of an entity is termed
immanent causation. It is crucial to distinguish between the iterated sequential
composite $\mathrm{Imm}_N \circ \mathrm{Imm}_N$ of the set-valued imminence mapping Imm_N and the
hierarchical composite $f \vdash g \vdash h$ among the standard single-valued mappings
f, g, and h entailed by the iterated imminence $h \in \mathrm{Imm}_N^2(f) =$
$\mathrm{Imm}_N \circ \mathrm{Imm}_N(f)$.

The three mappings f, g, h form a hierarchical chain

(13)

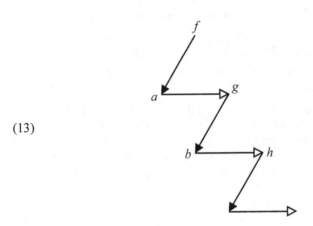

Note that only the existence of one such intermediary g is required: the mapping
f may functionally entail many more mappings in $\kappa(N)$, but none of these other
branches are obliged to immediately connect to h.

The iterated imminence $\mathrm{Imm}_N^2(f)$ may be interpreted as all the processes in
the natural system N that are reachable from the process f after two functional
entailment steps. This 'indirect imminence' is inherent in the imminence network
Imm_N and will turn out (anticipating Chapter 13) to be a relational
characterization of viruses.

9.10 Square Product Imminence   The square product $\mathrm{Imm}_N \circ \mathrm{Imm}_N : \kappa(N)$ $\to \kappa(N)$ is defined by

(14) $$\left(\mathrm{Imm}_N \circ \mathrm{Imm}_N\right)(f) = \bigcap_{g \in \mathrm{Imm}_N(f)} \mathrm{Imm}_N(g) \subset \kappa(N)$$

(*cf.* Definition 3.4.ii).

A mapping $h \in \left(\mathrm{Imm}_N \circ \mathrm{Imm}_N\right)(f)$, different from the iterated imminence $h \in \mathrm{Imm}_N^2(f)$, must be reachable from the mapping f after travelling on all two connected arrow-pairs initiating from f in the digraph Imm_N —all the mappings $g_1, g_2, \ldots, g_m \in \mathrm{Imm}_N(f)$ must entail h:

(15) $$f \vdash g_1 \vdash h, \quad f \vdash g_2 \vdash h, \quad \ldots, \quad f \vdash g_m \vdash h.$$

This means a relational diagram that contains the branching pattern

(16)

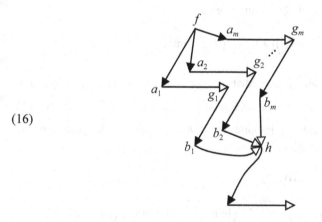

Let $\mathrm{dom}(f) = A$ and $h \in H(C,D)$. Then for each $i = 1,2,\ldots,m$, $a_i \in A$ and

(17) $$f(a_i) = g_i,$$

with the possibility that the m intermediary mappings g_i may all have different domain sets $B_i = \mathrm{dom}(g_i)$, whence the g_i s may belong to different hom-sets $H\left(B_i, H(C,D)\right) \subset \kappa(N)$. All that is required is that for each $i = 1,2,\ldots,m$,

(18) $$h \in H(C,D) \cap \mathrm{ran}(g_i) \neq \varnothing,$$

whence

(19) $$h \in \mathrm{Imm}_N(g_i) = \kappa(N) \cap \mathrm{ran}(g_i)$$

and therefore

(20) $$h \in \bigcap_{i=1}^{m} \mathrm{Imm}_N(g_i) = \bigcap_{g \in \mathrm{Imm}_N(f)} \mathrm{Imm}_N(g) = \left(\mathrm{Imm}_N \circ \mathrm{Imm}_N \right)(f).$$

Note the requirement (18) that $H(C,D) \cap \mathrm{ran}(g_i) \neq \varnothing$ does not necessarily imply that $g_i \in H(B_i, H(C,D))$. It may very well happen that for another $b_i' \in B_i$, $g_i(b_i') \notin H(C,D)$, as explained in Section 7.11 in connection with the imminence mapping Imm_C.

In contrast to immanent causation $\exists g$ in (12), now, square product imminence may symbolically be summarized as

(21) $$h \in \left(\mathrm{Imm}_N \circ \mathrm{Imm}_N \right)(f) \;\vdash\; \left(\forall g \in \mathrm{Imm}_N(f) : f \vdash g \vdash h \right).$$

The imposition on all intermediaries $g \in \mathrm{Imm}_N(f)$ entailed by $h \in \left(\mathrm{Imm}_N \circ \mathrm{Imm}_N \right)(f)$ is an inherent *redundancy*, a multiplicity of entailment paths that says something about the importance of h to require such protection and robustness to ensure its imminent repair.

9.11 Iterated Inverse   Trivially,

(22) $$h \in \mathrm{Imm}_N^2(f) \text{ iff } f \in \left(\mathrm{Imm}_N^2 \right)^{-1}(h) = \mathrm{Imm}_N^{-2}(h).$$

Corollary 3.7 says that

(23) $$f \in \mathrm{Imm}_N^{-2}(h) = \left(\mathrm{Imm}_N \circ \mathrm{Imm}_N \right)^{-1}(h) \quad \text{iff}$$
$$\mathrm{Imm}_N(f) \cap \mathrm{Imm}_N^{-1}(h) \neq \varnothing,$$

the nonempty intersection $\mathrm{Imm}_N(f) \cap \mathrm{Imm}_N^{-1}(h)$ entailing the existence of an intermediary mapping g in the hierarchical chain (13).

For the square product,

(24) $\quad h \in \left(\mathrm{Imm}_N \circ \mathrm{Imm}_N \right)(f) \quad \text{iff} \quad f \in \left(\mathrm{Imm}_N \circ \mathrm{Imm}_N \right)^{-1}(h),$

and Theorem 3.9 gives

(25) $\quad f \in \left(\mathrm{Imm}_N \circ \mathrm{Imm}_N \right)^{-1}(h) = \mathrm{Imm}_N^{+1} \left(\mathrm{Imm}_N^{-1}(h) \right),$

where the core (*cf.* Definition 2.16.ii) of $E \subset \kappa(N)$ by Imm_N is the set

(26) $\quad \mathrm{Imm}_N^{+1}(E) = \left\{ p \in \kappa(N) : \mathrm{Imm}_N(p) \subset E \right\}.$

Thus

(27) $\quad \begin{aligned} \left(\mathrm{Imm}_N \circ \mathrm{Imm}_N \right)^{-1}(h) &= \mathrm{Imm}_N^{+1} \left(\mathrm{Imm}_N^{-1}(h) \right) \\ &= \left\{ p \in \kappa(N) : \mathrm{Imm}_N(p) \subset \mathrm{Imm}_N^{-1}(h) \right\} \end{aligned}$

whence

(28) $\quad f \in \left(\mathrm{Imm}_N \circ \mathrm{Imm}_N \right)^{-1}(h) \quad \text{iff} \quad \mathrm{Imm}_N(f) \subset \mathrm{Imm}_N^{-1}(h),$

which is the first conclusion in Theorem 3.10 when specialized to $\mathrm{Imm}_N : \kappa(N) \multimap \kappa(N)$. The inclusion $\mathrm{Imm}_N(f) \subset \mathrm{Imm}_N^{-1}(h)$ precisely says that a mapping $f \in \left(\mathrm{Imm}_N \circ \mathrm{Imm}_N \right)^{-1}(h)$ (whence $h \in \left(\mathrm{Imm}_N \circ \mathrm{Imm}_N \right)(f)$) means that all the mappings in the imminence $\mathrm{Imm}_N(f)$ must entail h, as shown in the branching pattern (16).

Note that $\mathrm{Imm}_N(f) \subset \mathrm{Imm}_N^{-1}(h)$ does not preclude the possibility, in the network N, of the existence of an intermediary mapping $g \in \mathrm{Imm}_N^{-1}(h) \sim \mathrm{Imm}_N(f)$, a $g \in \kappa(N)$ that functionally entails h but is not functionally entailed by f. The other two conclusions of Theorem 3.10 when specialized to $\mathrm{Imm}_N : \kappa(N) \multimap \kappa(N)$, however, are

(29) $\quad f \in \left(\mathrm{Imm}_N^{-1} \circ \mathrm{Imm}_N^{-1} \right)(h) \quad \text{iff} \quad \mathrm{Imm}_N^{-1}(h) \subset \mathrm{Imm}_N(f),$

and

(30)
$$\mathrm{Imm}_N(f) = \mathrm{Imm}_N^{-1}(h) \quad \text{iff}$$
$$f \in \left(\mathrm{Imm}_N^{-1} \circ \mathrm{Imm}_N^{-1}\right)(h) \cap \left(\mathrm{Imm}_N \circ \mathrm{Imm}_N\right)^{-1}(h)$$

This last equivalence prescribes the condition under which the two-step square product imminence $h \in \left(\mathrm{Imm}_N \circ \mathrm{Imm}_N\right)(f)$ allows the two mappings f and h, two hierarchical levels apart, to determine each other uniquely, as well as to define the collection of intermediary mappings. In terms of the three mappings $\{f,g,h\}$ that functionally entail one another in cyclic permutation in the entailment diagram that is the very representation of the simplest (M,R)-system

(31)

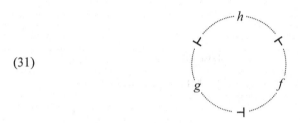

the equivalence (30) is a statement that says when two of a triplet $\{f,g,h\}$ of mappings may determine the third. This, of course, may be interpreted variously as *specificity*, *stability*, and *robustness* in interconnecting replication-repair-metabolism maps.

PART III
Interacting (M,R)-Systems

Mit dem Genius steht die Natur im ewigen Bunde,
Was der eine verspricht, leistet die andre gewiss.

[With genius Nature ever stands in solemn union still,
And ever what the One foretells the other shall fulfil.]

— Friedrich von Schiller (1795)
Columbus

This Part III deals with the relational biology of interactions.

I explicate the topology of the different modes of relational interactions of (M,R)-networks and then formulate in relational terms the ubiquitous biological interaction of symbiosis. Various relational interactions between (M,R)-networks also have realizations in pathophysiology, consequences in natural selection, implications on origins of life, and ramifications in virology. The possible reversals of the effects of these interactions become therapeutic models.

10
Connections

> Only connect! That was the whole of her sermon. Only connect
> the prose and the passion, and both will be exalted, and human
> love will be seen at its height. Live in fragments no longer.
> Only connect, and the beast and the monk, robbed of the
> isolation that is life to either, will die.

> — E. M. Forster (1910)
> *Howards End*
> Chapter 22

Relational Interactions

10.1 Six Modes of Nodal Connection   The relational diagrams of mappings
may interact: two mappings, with the appropriate domains and codomains, may
be connected at different common nodes. Since there are three nodes in the
relational diagram of each mapping, on account of symmetry, there are $\binom{3}{2} = 6$
possible modes of nodal connection:

A.H. Louie, *The Reflection of Life*, IFSR International Series on Systems Science
and Engineering 29, DOI 10.1007/978-1-4614-6928-5_10,
© Springer Science+Business Media New York 2013

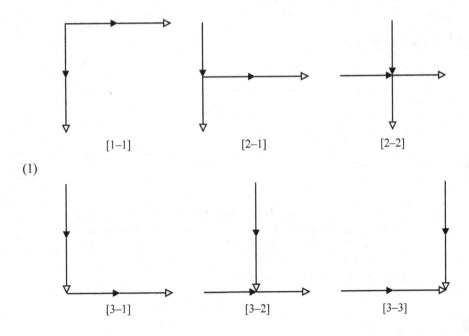

(1)

These six modes of connection model various aspects of relational interactions. These interactions are 'internal' when the two mappings are processes in the same system and are 'inter-network' when the two mappings belong to different entailment networks. We shall see their biological realizations in subsequent chapters.

10.2 Common Causes   The head of the hollow-headed arrow is identified with 'that which is entailed' (*cf.* Section 6.6). It is of most interest when 'that which is entailed' has a dual role: an alternate description as some entity in addition to being a final cause of its own mapping.   This happens when the relational diagrams of two mappings connect at the head of a hollow-headed arrow (as in the three networks [3–1], [3–2], and [3–3] in diagram (1)).

For an interaction to be a *composition*, one mapping must entail in the other an attribute that is subsequently relayed.   The connection [3–3]

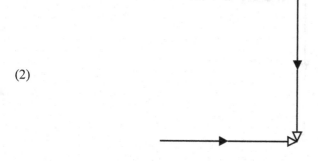

(2)

simply shows two mappings with a common codomain; while the two mappings have a common final cause (or at least the two outputs belong to the same set), the mappings nevertheless do not compose.

The connections [3–1] and [3–2] are two kinds of compositions that I have explicated in category-theoretic terms in Chapter 6: sequential composition [3–2] that is material entailment and hierarchical composition [3–1] that is functional entailment. I shall presently return to these two modes of 'dual entailment' modelled by two mappings in composition, in the context of interacting systems.

10.3 Resolutions   The connection [2–1] in diagram (1)

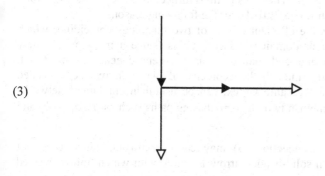

(3)

shows that the domain of one mapping consists of mappings; i.e. the material cause of one is the efficient cause of the other.

The two connections [1–1] and [2–2]

(4)

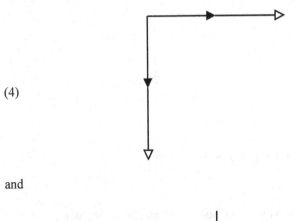

and

(5)

were disallowed in *ML*: 5.16. These restrictions turned out to be unnecessary (and have since been lifted, in Louie [2010]), for the following reasons:

Diagram (4) shows the efficient causes of two mappings coinciding, which implies that so must their domains and codomains (since a mapping uniquely determines its domain and codomain). This apparent degeneracy need not, however, be abandoned. Indeed, the geometry of (4), with two solid-headed arrows originating from the same vertex, can be useful in entailment networks when one wants to distinguish two element-chasing paths such as $f : a_1 \mapsto b_1$ and $f : a_2 \mapsto b_2$.

The 'crossed-path' connection (5) may cause confusion, since it is not immediately clear which solid-headed arrow is paired with which hollow-headed arrow. But the resolution required is simply that one must be careful in tracing paths. When circumstances warrant, the crossed-path may be unfolded into two disjoint paths thus

(6)

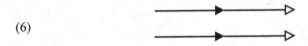

Note that the interactions (3) and (4) may also be similarly resolved, when necessary, without loss of entailment structure, into two disjoint paths (6).

The resolution of the degeneracy (4), the unfolding of the bifurcation, is equivalent to the vertex partition $v \to \{v_1, v_2\}$:

(7)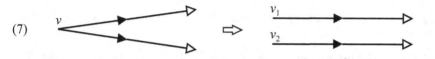

This process preserves the degree sum

(8) $$\varepsilon_o(v) = \varepsilon_o(v_1) + \varepsilon_o(v_2)$$

in the equivalent unfolded network (*cf. ML*: 6.8 for the explanation of the four degrees $\left(\varepsilon_i(v), \tau_i(v), \varepsilon_o(v), \tau_o(v)\right)$ of a vertex v in an entailment network G). In the unfolded network, the restrictions, for each $v \in G$,

(9) $$\varepsilon_o(v) = 0 \text{ or } 1$$

and

(10) $$\tau_i(v) \geq \varepsilon_o(v)$$

(conditions important in some of the topological arguments for traversability) still hold (*cf. ML*: 6.10 & 6.24).

Entailment Between Systems

Let **C** be a category with its collection **OC** of objects and its collection **AC** of morphisms (*cf.* Section 6.8). Consider two systems H and S in **C**, whence there are two sets of *efficient causes*, $\kappa(H)$ and $\kappa(S)$ (*cf.* Section 7.10), that are subsets of **AC**. Let $\upsilon(H)$ be the union of **C**-objects (sets) that are domains of **C**-morphisms (mappings) in H ; i.e.

(11) $$\upsilon(H) = \bigcup_{p \in \kappa(H)} \text{dom}(p),$$

which is the set of all *material causes* in H. ('Matter' in ancient Greek is the word ὕλη for 'wood'. Aristotle adapted it for matter in general, and by extension the material cause underlying a change, whence the symbol υ.) Note that $\upsilon(H)$ is a union of **C**-objects but it does not itself necessarily belong to $\mathcal{O}\mathbf{C}$ and that a is a material cause in H if and only if

(12) $\exists p \in \kappa(H) : a \in \mathrm{dom}(p) \in \mathcal{O}\mathbf{C}$ iff $a \in \upsilon(H)$.

10.4 Sequential Interaction   Let f and g be processes in H and S, respectively; i.e. $f \in \kappa(H)$ and $g \in \kappa(S)$. Then

(13) $\mathrm{dom}(f) \subset \upsilon(H)$ and $\mathrm{dom}(g) \subset \upsilon(S)$.

The connection [3–2] in diagram (1)

(14)

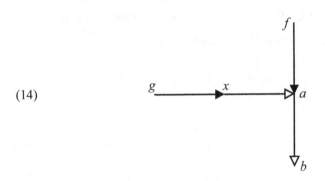

is *sequential composition* in the category **C**

(15) $\circ : (g, f) \mapsto f \circ g$

with corresponding entailment diagram

(16) $g \vdash a, \; f \vdash b \;\Rightarrow\; f \circ g \vdash b$

(*cf.* Section 6.10).

Recall that the binary operation $\circ$ is not defined for all pairs of **C**-morphisms. By definition, the codomain of the first argument must be identical to the domain of the second argument for the binary operation to proceed. In $f \circ g$, the requirement is

(17) $$\mathrm{cod}(g) = \mathrm{dom}(f),$$

which may be relaxed to

(18) $$\mathrm{ran}(g) \subset \mathrm{dom}(f),$$

as explained in Section 6.10. The necessary and sufficient condition (18) for $f \circ g \in \mathfrak{RC}$ may be further weakened, if one is concerned not about the 'complete' composite mapping $f \circ g : \mathrm{dom}(g) \to \mathrm{cod}(f)$ but just specific 'relays' $f \circ g : x \mapsto g(x) \mapsto f(g(x))$ as in diagram (14). Given $g \in \mathfrak{RC}$, let

(19) $$X_f = \{x \in \mathrm{dom}(g) : g(x) \in \mathrm{dom}(f)\}.$$

For $x \in X_f \subset \mathrm{dom}(g)$, $g(x) \in \mathrm{dom}(f) \cap \mathrm{ran}(g) \neq \varnothing$, whence f may accept $g(x)$ as input and proceed to materially entail $f \circ g(x) = f(g(x))$. The nonempty intersection

(20) $$\mathrm{dom}(f) \cap \mathrm{ran}(g) \neq \varnothing$$

is a prerequisite for the inclusion requirement (18) of the 'complete' composite mapping $f \circ g : \mathrm{dom}(g) \to \mathrm{cod}(f)$, but (20) by itself only entails that $f \circ g$ may be 'partially' defined, as the *restriction* $f \circ g|_{X_f} : X_f \to \mathrm{cod}(f)$. (It may happen that for some $y \in \mathrm{dom}(g)$, $g(y) \notin \mathrm{dom}(f)$, whence $f \circ g(y) = f(g(y))$ is not defined.)

10.5 Metabolism Bundle   Now, define a set-valued mapping $M_{S \to H} : \kappa(S) \multimap \upsilon(H)$ by, for a mapping $q \in \kappa(S)$,

(21) $$M_{S \to H}(q) = \upsilon(H) \cap \mathrm{ran}(q).$$

For $f \in \kappa(H)$ and $g \in \kappa(S)$, the partial sequential composition $f \circ g|_{X_f} : X_f \to \mathrm{cod}(f)$, as explained in the previous paragraph, may be defined if

(22) $$M_{S \to H}(g) = \upsilon(H) \cap \mathrm{ran}(g) \supset \mathrm{dom}(f) \cap \mathrm{ran}(g) \neq \varnothing.$$

Stated otherwise, $M_{S \to H}(g)$ contains all the entities in the system H that may be materially entailed by the process $g \in \kappa(S)$ of the system S. Material entailment is *metabolism* in its most general sense, whence $M_{S \to H}(g)$ may be considered the *metabolism effect* of the interaction $g : S \to H$.

The 'product over $\mathbf{OC}$', the domain $\mathfrak{A}\mathbf{C} \times_{\mathbf{OC}} \mathfrak{A}\mathbf{C}$ of the binary operation $\circ$ of sequential composition on $\mathfrak{A}\mathbf{C}$, is

(23) $$\mathfrak{A}\mathbf{C} \times_{\mathbf{OC}} \mathfrak{A}\mathbf{C} = \left\{ (q, p) \in \mathfrak{A}\mathbf{C} \times \mathfrak{A}\mathbf{C} : \text{dom}(p) = \text{cod}(q) \right\}$$

(*cf.* Section 6.10); if $(q, p) \in \mathfrak{A}\mathbf{C} \times_{\mathbf{OC}} \mathfrak{A}\mathbf{C}$, then $p \circ q \in \mathfrak{A}\mathbf{C}$. For partial sequential composition, one may expand $\mathfrak{A}\mathbf{C} \times_{\mathbf{OC}} \mathfrak{A}\mathbf{C}$ to a larger subset of $\mathfrak{A}\mathbf{C} \times \mathfrak{A}\mathbf{C}$:

(24) $$\text{Met}_{\mathbf{C}} = \left\{ (q, p) \in \mathfrak{A}\mathbf{C} \times \mathfrak{A}\mathbf{C} : \text{dom}(p) \cap \text{ran}(q) \neq \varnothing \right\};$$

if $(q, p) \in \text{Met}_{\mathbf{C}}$, then $p \circ q|_{X_p} : X_p \to \text{cod}(p)$ may be defined (but $p \circ q|_{X_p}$ may not be in the existing collection $\mathfrak{A}\mathbf{C}$ of **C**-morphisms). The subset $\text{Met}_{\mathbf{C}} \subset \mathfrak{A}\mathbf{C} \times \mathfrak{A}\mathbf{C}$ is the domain on which the 'metabolism effect' in **C** may proceed in the relay $p \circ q|_{X_p} : x \mapsto q(x) \mapsto p(q(x))$, hence the expression ' Met ' as the symbol. One may also note that $\text{Met}_{\mathbf{C}}$, as a subset of $\mathfrak{A}\mathbf{C} \times \mathfrak{A}\mathbf{C}$ (whence a relation on $\mathfrak{A}\mathbf{C}$), has an alternate representation as a set-valued mapping:

(25) $$\text{Met}_{\mathbf{C}} : \mathfrak{A}\mathbf{C} \multimap \mathfrak{A}\mathbf{C} .$$

The domain $\text{Met}_{\mathbf{C}}$ of the binary operation $\circ$ of partial sequential composition in **C** is a *bundle* in the category-theoretic sense. Let

(26) $$\pi_1, \pi_2 : \text{Met}_{\mathbf{C}} \to \mathfrak{A}\mathbf{C}$$

be the canonical projection maps (onto the first and second components, respectively). Then for $g \in \mathfrak{A}\mathbf{C}$, the inverse image $\pi_1^{-1}(g)$ (generalized '*fibre over g*') is the set

(27) $$\pi_1^{-1}(g) = \left\{ (g, p) : p \in \mathfrak{A}\mathbf{C}, \ \text{dom}(p) \cap \text{ran}(g) \neq \varnothing \right\} \subset \text{Met}_{\mathbf{C}},$$

and its projected image onto the second component (the g-dependent 'fibre' itself)

(28)
$$\pi_2\left(\pi_1^{-1}(g)\right) = \left\{p \in \mathfrak{R}\mathbf{C} : \mathrm{dom}(p) \cap \mathrm{ran}(g) \neq \varnothing\right\}$$

contains all the **C**-morphisms p that can partially compose with g in $p \circ g|_{X_p}$.

When 'regionally' restricted to two interacting systems H and S in **C**, the *metabolism bundle* $\mathrm{Met}_{S \to H} : \kappa(S) \multimap \kappa(H)$ takes the form

(29)
$$\mathrm{Met}_{S \to H} = \left\{(q, p) \in \kappa(S) \times \kappa(H) : \mathrm{dom}(p) \cap \mathrm{ran}(q) \neq \varnothing\right\};$$

if $(q, p) \in \mathrm{Met}_{S \to H}$, then $p \circ q|_{X_p} : X_p \to \mathrm{cod}(p)$ may be defined (but $p \circ q|_{X_p}$ may not be in the existing collections $\kappa(H)$ or $\kappa(S)$ of processes). The projectors are

(30)
$$\pi_1 : \mathrm{Met}_{S \to H} \to \kappa(S), \quad \pi_2 : \mathrm{Met}_{S \to H} \to \kappa(H).$$

For $g \in \kappa(S)$, one has

(31)
$$\pi_2\left(\pi_1^{-1}(g)\right) = \left\{p \in \kappa(H) : \mathrm{dom}(p) \cap \mathrm{ran}(g) \neq \varnothing\right\},$$

and

(32)
$$M_{S \to H}(g) = \bigcup_{p \in \pi_2\left(\pi_1^{-1}(g)\right)} \mathrm{dom}(p);$$

i.e. the metabolism effect $M_{S \to H}(g)$ of the interaction $g : S \to H$ is the union of the domains of all the mappings in the g-'fibre' of the metabolism bundle $\mathrm{Met}_{S \to H}$.

10.6 Hierarchical Interaction The connection [3–1] in diagram (1)

(33)

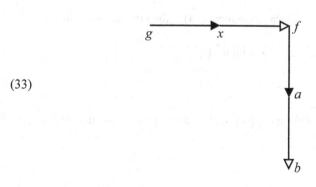

is the hierarchical composition with the corresponding composition of entailment diagrams

(34) $$g \vdash f, \; f \vdash b \;\Rightarrow\; g \vdash f \vdash b$$

(*cf.* Section 6.13). $f \in \kappa(H)$ is functionally entailed by $g \in \kappa(S)$ if and only if

(35) $$f \in \mathrm{ran}(g).$$

Now, define a set-valued mapping $F_{S \to H} : \kappa(S) \multimap \kappa(H)$ by, for a mapping $q \in \kappa(S)$,

(36) $$F_{S \to H}(q) = \kappa(H) \cap \mathrm{ran}(q).$$

Hierarchical composition $g \vdash f$ may be defined for $f \in \kappa(H)$ and $g \in \kappa(S)$ if and only if

(37) $$f \in \kappa(H) \cap \mathrm{ran}(g) = F_{S \to H}(g) \neq \varnothing.$$

Stated otherwise, $F_{S \to H}(g)$ contains all the entities in the system H that may be functionally entailed by the process $g \in \kappa(S)$ of the system S. The set-valued mapping $F_{S \to H}$ may, therefore, be considered the *imminence of S on H* , i.e. *inter-network imminence.* Functional entailment is *repair* in its most general sense, whence $F_{S \to H}(g)$ may be considered the *repair effect* of the interaction $g : S \to H$.

When the two systems involved are the same, then the mapping $F_{N \to N}$ of 'self-repair effects' is the imminence mapping of the system; viz.,

(38) $$F_{N \to N} = \mathrm{Imm}_N : \kappa(N) \multimap \kappa(N)$$

(*cf.* Section 9.1). The hierarchical entailment (34) is then the iterated imminence:

(39) $$b \in \mathrm{Imm}_N^2(g)$$

(*cf.* Section 9.9).

10.7 Range Partition   For $g \in \kappa(S)$ and $x \in \mathrm{dom}(g)$, consider the output $g(x) \in \mathrm{ran}(g).$

If

(40) $$g(x) \in M_{S \to H}(g) = \upsilon(H) \cap \mathrm{ran}(g),$$

then

(41) $$\exists f \in \kappa(H) : g(x) \in \mathrm{dom}(f) \subset \upsilon(H),$$

and the (partial) sequential composition $f \circ g$ may be defined at x, with

(42) $$(f \circ g)(x) = f(g(x)).$$

The process $g \in \kappa(S)$ of the system S entails the material cause $g(x)$ that is the input of the process $f \in \kappa(H)$ of the system H.

If

(43) $$g(x) \in F_{S \to H}(g) = \kappa(H) \cap \mathrm{ran}(g),$$

then

(44) $$\exists f \in \kappa(H) : g \vdash f,$$

and the hierarchical composition $g \vdash f$ may be defined, with

(45) $$g(x) = f.$$

The process $g \in \kappa(S)$ of the system S entails the efficient cause $g(x)$ that is the process $f \in \kappa(H)$ itself of the system H.

While the membership conditions (40) and (43) both serve to connect the two entailment networks S and H, they are not the only two possibilities. The inclusion

(46) $$M_{S \to H}(g) \cup F_{S \to H}(g) \subset \mathrm{ran}(g)$$

may be proper: the interaction $g : S \to H$ may comprise more than its metabolism and repair effects. This is to say, it may very well happen that for some $x \in \mathrm{dom}(g)$, $g(x) \in \mathrm{ran}(g)$ is disjoint from the system H, with

(47) $$M_{S \to H}(g) = F_{S \to H}(g) = \varnothing.$$

One may also note it is feasible that for some $x \in \text{dom}(g)$,

(48) $$g(x) \in M_{S \to H}(g) \cap F_{S \to H}(g),$$

whence

(49) $$M_{S \to H}(g) \cap F_{S \to H}(g) \neq \varnothing.$$

This happens when $g \in \kappa(S)$ supplies the material cause $g(x)$ of one process $f_1 \in \kappa(H)$, and $g(x)$ is simultaneously the efficient cause of another process $f_2 \in \kappa(H)$ itself:

(50)

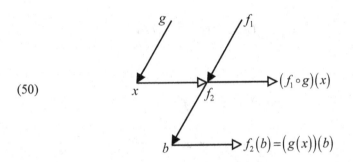

This relational diagram is an example of multiple connections among mappings, my next topic.

Multiple Connections and Unfolding

10.8 Unfolding (M,R)-Network   It is possible that the relational interaction of two mappings consists of connections at more than one common node.   The relational diagram may be resolved by unfolding into single connections for entailment analysis.   For example, the relational diagram of the simplest metabolism–repair connection (*cf. ML*: 12.13)

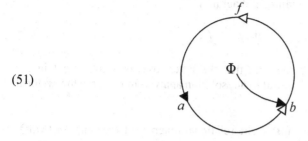

(51)

may be unfolded into the hierarchical composition

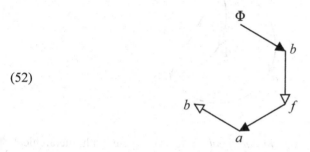

(52)

while preserving the entailment

(53) $\Phi \vdash f \vdash b$.

Note that the phrase 'while preserving the entailment' is important here. This is because diagram (51) may also be unfolded into the sequential composition

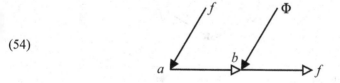

(54)

which abbreviates to the relational diagram

(55)

whence the corresponding entailment diagram is

(56) $\Phi \circ f \vdash f$.

Comparing (53) with (56), one sees that the latter loses one entailment in the process. Thus, one must be careful in a resolution analysis to preserve hierarchical compositions.

10.9 Unfolding (M,R)-Systems   The relational diagram of the simplest (M,R)-system (*cf. ML*: 12.14) is

(57)

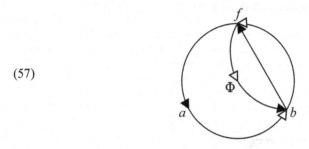

It may be unfolded into the *hierarchical cycle* (i.e. cycle with hierarchical compositions)

(58)

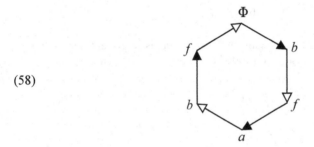

with the cyclic entailment

(59)

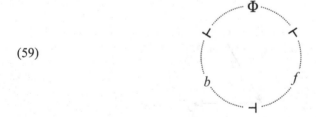

For each of the other two alternate encodings of replication (*cf. ML*: 12.15 & 12.16), the (M,R)-system also unfolds into a similar hierarchical cycle

(60)

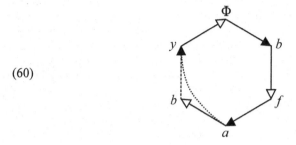

with an appropriate choice of y, the material cause of replication. (For the second class of (M,R)-systems, $y = b$; for the third class of (M,R)-systems, $y = a$; *cf. ML*: Chapter 12.)

11
Symbiosis

Great fleas have little fleas upon their backs to bite 'em,
And little fleas have lesser fleas, and so *ad infinitum*.
And the great fleas themselves, in turn, have greater fleas to go on;
While these again have greater still, and greater still, and so on.

— Augustus de Morgan (1872)
A Budget of Paradoxes

Natural Philosophy of Symbiosis

11.1 Definition   *Symbiosis* (Greek σύν with, βίωσις living; literally 'living together') is the close and often long-term biological interaction between different species, thence called *symbionts*.

In this definition, the term 'species' is used in a very general sense to mean 'a class of entities having some common characteristics'. A 'species', in other words, is simply a group containing individuals that agree in some common attributes and are called by a common name. The usage includes its biological sense of a group of living organisms (consisting, more specifically, of related similar individuals capable of gene exchange or interbreeding). In symbiosis, since the connection between entailment networks is 'biological interaction', one normally assumes that at least one of the symbiont 'species' is a living system.

11.2 Effect   A symbiotic relationship between individuals of two different species may be categorized by effect as mutualistic, commensal, or parasitic: *mutualism* is a relationship where both derive a benefit, *commensalism* is a relationship where one benefits and the other is neither significantly harmed nor helped, and *parasitism* is a relationship in which one benefits, while the other is harmed.

The relationship is often not symmetric, even in mutualism. It is usually obvious that one of the two symbionts is the *host*, in the sense that it harbours the other and provides substrate, shelter, or nourishment. With the usage of 'host',

A.H. Louie, *The Reflection of Life*, IFSR International Series on Systems Science and Engineering 29, DOI 10.1007/978-1-4614-6928-5_11,
© Springer Science+Business Media New York 2013

'symbiont' whence refers to the 'non-host partner' in the symbiosis. Thus, in context, one may write, for example, "two symbionts" in a symbiotic relationship, as well as the conjunction "host–symbiont" (especially "host–parasite").

A stereotypical example of mutualism is the relationship between the clownfish and the sea anemone. The clownfish feeds on small invertebrates which may otherwise harm the sea anemone, excrement from the clownfish provides nutrients to the sea anemone, and the territorial clownfish fends off anemone-eating predators. Reciprocally, in addition to providing ambient food for the clownfish, the sea anemone also offers a protective dwelling armed with its stinging tentacles, the clownfish being immune to the nematocysts and potent toxins of its host. Barnacles are sedentary crustaceans that as habitat attach themselves permanently to a substrate, including another organism (e.g. shellfish, whale), with the commensalism leaving the substrate organism unharmed. Almost all organisms have internal and external parasites, for example, louse, hookworm, dodder, and mould. *Infection* is a detrimental form of parasitism, in which the host is colonized by a foreign (usually microscopic) species.

11.3 Topology Symbiosis may alternatively be categorized by topology. *Endosymbiosis* is a symbiotic relationship in which one symbiont lives within the tissues of the other. An example is the nitrogen-fixing bacteria that live in nodules on legume roots. Lichen is an endosymbiotic union between a fungus and a photosynthetic alga. The endosymbiotic theory of cell evolution—which certain organelles (such as mitochondria and chloroplasts) of eukaryotic cells are originally separate prokaryotic organisms taken inside as endosymbionts—has been generally accepted. Contrastively, *ectosymbiosis* is a symbiotic relationship in which the symbiont lives on the surface (which includes the outer body surface as well as the inner surface such as that of the digestive tract or the ducts of exocrine glands) of the host. Many herbivores, for example, have gut fauna that help them digest plant matter.

11.4 Necessity Symbiosis may also be categorized by necessity as either obligate or facultative: an *obligate* relationship is necessary for the survival of at least one of the organisms involved, while a *facultative* relationship may be beneficial but not essential for survival of the organisms. An extreme example of obligate mutualism is between the tube worms and symbiotic bacteria that live at hydrothermal vents and cold seeps: the worm has no digestive tract and is completely dependent on its internal symbionts for nutrition. Most myrmecophilous associations between ants and a variety of other organisms (plants, arthropods, and fungi) are facultative mutualisms.

Viruses are obligate intracellular parasites—viral genomes only function after they replicate in a cell. Viruses are not alive: they are collections of chemicals and, by themselves, do not reproduce; a cellular host is needed. While symbiosis (hence its subcategory parasitism) almost always concerns organisms (i.e. living systems), Definition 11.1, in general terms of interacting species, allows the inclusion of viruses.

Remember Humpty Dumpty: this broad definition of symbiosis that I am using—simply as "the living together of closely interacting species" —is, however, not universal. Its usage is often more narrow and describes only those relationships from which both organisms benefit, that is, synonymous with mutualism. See Wilkinson [2001] for a succinct discussion on this very topic.

Relational Symbiosis

11.5 An Example   Consider two types of cells, denoted by H and S. Suppose S requires a metabolite v as input in a metabolic process $g_1 : v \mapsto w$, but it cannot produce v for itself. If H produces v in one of its own metabolic processes $f_1 : u \mapsto v$, then it is clear that the interaction between H and S

(1)

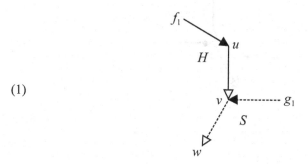

will be beneficial to S. (I have, for clarity, shown the arrows of the relational diagram of the processor $g_1 \in \kappa(S)$ with dashed lines to distinguish them from those of the processor $f_1 \in \kappa(H)$.) Remember the processes f_1 and g_1 are but single components of large entailment networks that are the cells H and S, respectively; diagram (1) simply isolates them to show their interaction. This connection makes S either a commensal or parasitic symbiont, depending on the effect of sharing metabolite v on the host H.

If, in addition, H requires a metabolite y that it itself cannot produce but S can, then their interaction may be a mutualism, a reciprocal sharing of the metabolites v and y:

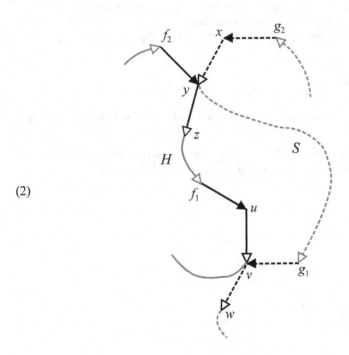

(2)

With this simple example as illustration, I define symbiosis in relational–biological terms thus:

11.6 Definition   *Symbiosis* is the close and often long-term relational interaction between different (M,R)-networks.

11.7 Modes of Relational Interaction   Henceforth, I shall use the generic hierarchical cycle

(3)

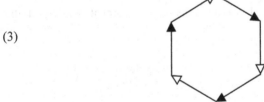

to symbolically represent a living system in general and the host H in a symbiotic relationship in particular. But one must keep in mind, however, that the entailment network of an organism is in general a far more complicated relational diagram consisting of a large number of interconnected arrows. When appropriate, the nodes may be labelled thus:

(4)

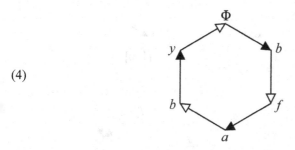

to denote the various processes $f, \Phi, b, a \in \kappa(H)$, that is, metabolism–repair–replication mappings of the (M,R)-system H (where the material cause of replication $y = f, b, a$, respectively, for the three classes of (M,R)-systems; *cf. ML*: Chapter 12).

I shall use

(5)

to denote a relational process $g \in \kappa(S)$ in a symbiont S. Of course, just like that of the host, the entailment network of the symbiont is a complex relational diagram. But the interactions between diagrams (3) and (5) are sufficient to show all the modes of symbiotic relationships.

There are $3 \times 2 = 6$ possible modes of nodal connection between host and symbiont:

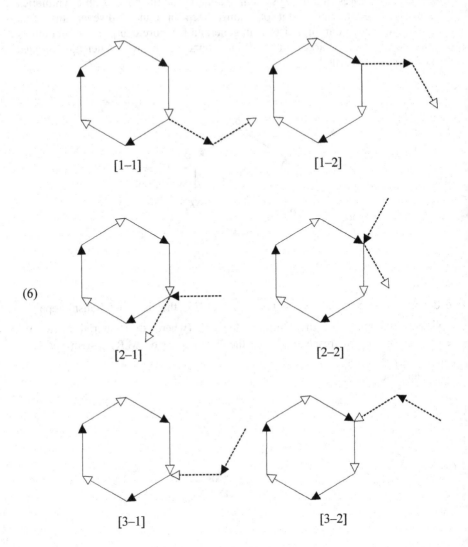

(6)

[1–1]   [1–2]

[2–1]   [2–2]

[3–1]   [3–2]

Symbiosis of (M,R)-Systems

11.8 Metabolism Symbiosis   The symbiotic relationship ([2–1] of diagram (6))

(7)

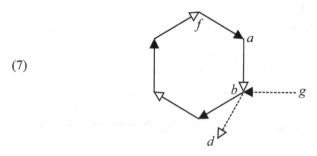

is the sharing (or, in the parasitic case, the siphoning off) of the metabolite b, a scenario that I discussed in the illustrative example of Section 11.5. The symbiosis is represented by the sequential composition $g \circ f$,

(8)
$$(g \circ f)(a) = g(f(a)) = g(b) = d,$$

with corresponding entailment diagram

(9)
$$f \vdash b, \quad g \vdash d \;\Rightarrow\; g \circ f \vdash d.$$

The material entailment set-valued mapping $M_{H \to S} : \kappa(H) \multimap \upsilon(S)$ (*cf.* Section 10.5) is such that

(10)
$$b = f(a) \in M_{H \to S}(f) = \upsilon(S) \cap \mathrm{ran}(f).$$

The material cause of $g \in \kappa(S)$,

(11)
$$b = f(a) \in \mathrm{dom}(g) \subset \upsilon(S),$$

is entailed by the process $f \in \kappa(H)$.

   That the connection between H and S in [2–1] of diagram (6) occurs at a 'relay node' of functional entailment (i.e. a node where the head of a hollow-headed arrow joins the tail of a solid-headed arrow) means the symbiotic relationship may dually be realized thus:

(12)

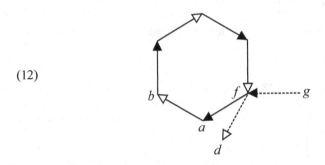

In this arrangement, the efficient cause $f \in \kappa(H)$ is used by $g \in \kappa(S)$ as material cause:

(13) $$f \in \operatorname{dom}(g) \subset \upsilon(S).$$

Biologically, this happens, for example, when a functional enzyme in H is metabolized as a nutrient protein by S.

The symbiotic relationship ([2–2] of diagram (6))

(14)

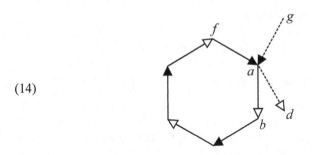

is similar, in this case the sharing of the metabolite a (or the competing for the resource a). Here, the relation between the mappings $f \in \kappa(H)$ and $g \in \kappa(S)$ is

(15) $$a \in \operatorname{dom}(f) \cap \operatorname{dom}(g) \subset \upsilon(H) \cap \upsilon(S) \neq \varnothing.$$

The connections (12) and (14) are not compositions. In diagrams (7), (12), and (14), the processor g of the symbiont S simply takes its material cause from the host, whence for S it is strictly a metabolic arrangement.

11.9 Repair Symbiosis   The interaction modes when the symbiont appropriates an efficient cause from the host are represented by

(16)

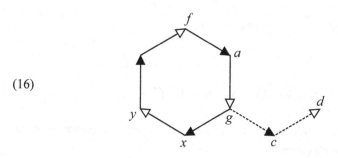

and

(17)

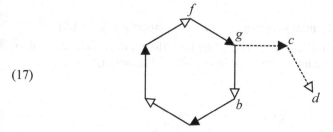

([1–1] and [1–2], respectively, of diagram (6)). In each of these two cases, the host resource shared with the symbiont is used by S in turn as its own processor.

In connections (16), the common resource g may become a commodity for competition between systems H and S. When

(18) $$g \in \kappa(H) \cap \kappa(S),$$

the symbiont process $g : c \mapsto d$ in $\kappa(S)$ may overwhelm in $\kappa(H)$ the original host processes $g : x \mapsto y$. Interaction (16) is dually the hierarchical composition

(19) $$f \vdash g, \; g \vdash d \;\Rightarrow\; f \vdash g \vdash d.$$

The corresponding functional entailment set-valued mapping $F_{H \to S} : \kappa(H) \multimap \kappa(S)$ (*cf.* Section 10.6) is such that

(20) $$g \in F_{H \to S}(f) = \kappa(S) \cap \mathrm{ran}(f) \neq \varnothing.$$

Interaction (17) says the relation between the mappings $f \in \kappa(H)$ and $g \in \kappa(S)$ is

(21) $$g \in \mathrm{dom}(f),$$

so

(22) $$\kappa(S) \cap \upsilon(H) \supset \kappa(S) \cap \mathrm{dom}(f) \neq \varnothing.$$

In this connection, the symbiont process $g : c \mapsto d$ may deprive the host process $f : g \mapsto b$ of its material cause.

Infection

11.10 Bacterial and Fungal Infection   When the processor $g \in \kappa(S)$ of the symbiont supplies its final cause (effect) to the host, the result is often harmful to the latter. In a metabolic interaction, one has ([3–2] of diagram (6))

(23)

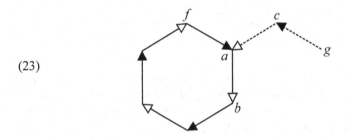

This is sequential composition $f \circ g$ with entailment diagram

(24) $$g \vdash a, \quad f \vdash b \quad \Rightarrow \quad f \circ g \vdash b.$$

The sequential composition is made possible by

(25) $$g(c) \in \mathrm{dom}(f) \subset \upsilon(H),$$

whence

(26) $$g(c) \in M_{S \to H}(g) = \upsilon(H) \cap \mathrm{ran}(g)$$

(with the material entailment set-valued mapping $M_{S \to H} : \kappa(S) \sqsubset \upsilon(H)$ as defined in Section 10.5).

In this interaction, the metabolic process $f \in \kappa(H)$ of the host, instead of its original mapping $f : a \mapsto b$, now becomes the composite $f \circ g : c \mapsto g(c)$ $\mapsto f(g(c))$. Stated otherwise, the material cause a of f is replaced by $g(c)$, so the original output $b = f(a)$ in the hierarchical cycle of the (M,R)-system of the host is then replaced by the different metabolite $b' = f(g(c))$. This foreign material may be realized as a toxin or an infective agent. Most administrations of defensive and predatory biotoxins and bacterial and fungal infections of organisms are relational–biological interactions of this mode.

When the substituted node of the host is an efficient cause rather than a material cause, the infection is often more devastating. The interaction ([3–1] of diagram (6)) may be realized as

(27)

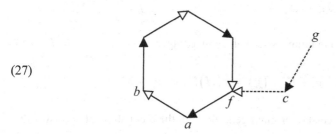

This is the hierarchical composition:

(28) $$g \vdash f, \quad f \vdash b \quad \Rightarrow \quad g \vdash f \vdash b,$$

whence the functional entailment set-valued mapping $F_{S \to H} : \kappa(S) \sqsubset \kappa(H)$ (*cf.* Section 10.6) with

(29) $$f \in F_{S \to H}(g) = \kappa(H) \cap \mathrm{ran}(g) \neq \varnothing .$$

The interaction displaces the original enzyme f in the (M,R)-system H with an antigen $f' = g(c)$, and the surrogate enzyme catalyses a different metabolic process $f' : a' \mapsto b'$ instead of the original $f : a \mapsto b$. Since now a new efficient cause is in place (and not, as in diagram (23), just a new material cause which may be of limited supply), the host is infected to produce its own infective material. Some biotoxins, bacteria, and fungi are infectious agents with relational–biological interactions of this mode, on a higher hierarchical level.

11.11 Viral Infection   The interaction that is [3–1] of diagram (6) may be realized on an even higher hierarchical level still. The genetic interaction

(30)

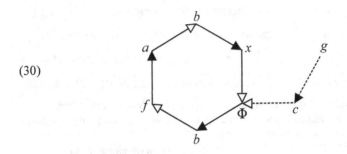

is the hierarchical composition

(31) $g \vdash \Phi, \; \Phi \vdash f \;\Rightarrow\; g \vdash \Phi \vdash f,$

whence the functional entailment set-valued mapping $F_{S \to H} : \kappa(S) \multimap \kappa(H)$ with

(32) $\Phi \in F_{S \to H}(g) = \kappa(H) \cap \mathrm{ran}(g) \neq \varnothing.$

This interaction replaces the original gene Φ with the rebel $\Phi' = g(c)$. Now the host executes a new genetic instruction $\Phi' : b' \mapsto f'$ instead of the old $\Phi : b \mapsto f$, thus producing new copies of the antigenic enzyme f' for further metabolic devastation. This is the mode of infection of the obligate intracellular molecular parasites that are viruses.

   I shall in subsequent chapters further explicate infection from the standpoint of relational biology.

11.12 Ubiquity of Symbiosis   Biology is a subject concerned with organization of relations. This fact is epitomized in relational biology by our definition of an organism as the realization of an (M,R)-system, a closed-to-efficient-causation entailment network in a nutshell (*cf.* Postulate of Life, 8.30 & *ML*: 11.28). The biological interaction of organisms is, therefore, the realization of the relational interaction of (M,R)-systems.

   This chapter has shown how various aspects of symbiosis are modelled in relational terms, as a result of single-node connections between host and symbiont, and the entailments of our theory are already consequential. The power of our approach that is relational biology is evident: each of the six modes of interaction (as listed in diagram (6) in Section 11.7) has a realization in symbiotic terms.

Symbiosis—in our general sense of commensalism and parasitism in addition to mutualism—is ubiquitous in the living world and is, indeed, an essential aspect of life itself. It may even be said that competition and symbiosis are the two driving forces of the biosphere. The importance of symbiosis in evolutionary innovation is evident when one understands its role in the determination of phenotypes and genotypes, as illustrated by the various entailment modes between symbiont and host. Darwin was fascinated by examples of 'co-evolution', in which organisms had adapted to each other and had to have evolved together (such as a flower with deep-buried nectar and a moth with a tongue long enough to reach it). One may even extend the definition of 'organism' to be more than single genetic entities, and include symbiotic units. But in relational–biological terms, this generalization is already made: a union of interacting (M,R)-systems (or better, their *join* in the *lattice* of (M,R)-systems; *cf. ML*: 2.1 & 7.28) is itself an (M,R)-system.

12
Pathophysiology

Les effets varient en raison des conditions qui les manifestent, mais les lois ne varient pas. L'état physiologique et l'état pathologique sont régis par les mêmes forces, et ils ne diffèrent que par les conditions particulières dans lesquelles la loi vitale se manifeste.

[Effects vary with the conditions that bring them to pass, but laws do not vary. Physiological and pathological states are ruled by the same forces, and they differ only because of the specific conditions under which the vital laws manifest themselves.]

— Claude Bernard (1865)
Introduction à l'étude
de la médecine expérimentale
Chapitre I

From Symbiosis to Pathophysiology

12.1 Definition *Pathophysiology* is the physiology of abnormality.

As a medical subject, pathophysiology is the study of the changes in biological functions entailed by prodromes, syndromes, or diseases. In biology, the science in which exceptions are the rule, it is important to note what is abnormal can be beneficial. Mutagenesis may lead to cancer and other heritable diseases, but it is also the driving force of evolution. In this chapter we shall explore abnormality in relational–biological terms.

12.2 Interacting Entailment Networks The generic hierarchical cycle that is an (M,R)-system distilled into its very essence

A.H. Louie, *The Reflection of Life*, IFSR International Series on Systems Science and Engineering 29, DOI 10.1007/978-1-4614-6928-5_12,
© Springer Science+Business Media New York 2013

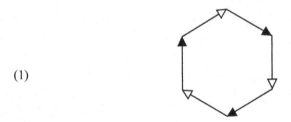

(1)

symbolically represents a living system. Let an additional relational process (e.g.
a perturbation of an internal process, an external influence) be denoted

(2)

In the previous chapter, diagrams (1) and (2) depicted, respectively, the host H
and the symbiont S of a symbiotic relationship. The same caveat applies here: the
entailment network of an organism is, of course, a far more complicated relational
diagram consisting of a large number of interconnected arrows, but the
interactions between these two diagrams are sufficient for my illustrative purposes.
I shall continue in what follows to use H and S when referring to the living
system and the additional process, although the two systems may not in general
necessarily be in symbiosis (and indeed, S may simply be a subsystem of H,
partitioned for the study of its perturbation effects).

12.3 Relational Pathophysiology   We have learned in the previous chapter that
when the processor of the symbiont supplies its final cause to the host, the effect,
realized as either *toxin* or *infection*, is often harmful to the latter.
        In a metabolic interaction, one has

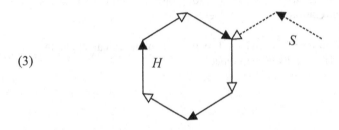

(3)

When the symbiont functionally entails a process of the host, as in

(4)

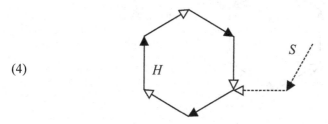

the toxicity or infection is often more devastating. These two modes of interaction are realized by toxin-producing organisms and by various infectious agents such as bacteria, fungi, and viruses (and prions).

Formulated in terms of set-valued mappings, diagram (3) is the metabolism effect

$$\text{(5)} \qquad M_{S \to H} : \kappa(S) \multimap \upsilon(H)$$

(*cf.* Sections 10.5 and 11.10) and diagram (4) is the repair effect

$$\text{(6)} \qquad F_{S \to H} : \kappa(S) \multimap \kappa(H)$$

(*cf.* Sections 10.6, 11.10, and 11.11).

From the standpoint of relational biology, pathophysiology is, then, the science of decodings and realizations of diagrams (3) and (4) and the study of consequences inherent in the set-valued mappings (5) and (6). In this chapter, I shall elaborate on these two relational diagrams and two set-valued mappings and explore what pathophysiological phenomena, in addition to infection, the relational interactions may encode and entail.

Entailment of Metabolism

12.4 Material Entailment as Metabolic Interaction

Let S contain a process g that supplies a new material cause to the metabolic process f of H.

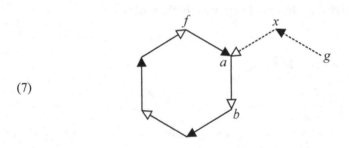

(7)

Stated otherwise, the material cause a of f is replaced by $a' = g(x)$; this says $a' \in M_{S \to H}(g)$. In essence, the metabolic process $f \in \kappa(H)$ is replaced by the sequential composite $f \circ g$.

If $f(a') = f(a)$, then the perturbation entailed by g, the variation $f \to f \circ g$, has no effect. Within the (M,R)-system H, the original entailment $f : a \mapsto b$ becomes $f : a' \mapsto b$, but since the new material cause a' is metabolized by f into the same final cause b, H carries on as before.

If $f(a') \neq f(a)$, however, the original output $b = f(a)$ in the hierarchical cycle of the organism H is then replaced by the different metabolite $b' = f(a')$. The foreign materials a' and b' may be considered infective agents. Most bacterial and fungal infections of organisms are relational–biological interactions of this mode, supplying 'foreign materials' into the metabolic network of the host. The immune response and the administration of antibiotics are sample processes that aim to terminate the interaction imposed by S on H. A 'treatment' may proceed in various ways. The most direct process is eliminating g altogether (i.e. the 'termination' of S). It may also work by detaching H and S, thence the two systems are no longer relationally connected (i.e. the 'removal' of S). A third alternative is to modify g sufficiently into g' (i.e. the 'alteration' of S, by the imposition therein of the functional entailment $\vdash g'$), so that the new material cause $a'' = g'(x)$ supplied to f is such that $f(a'') = f(a)$. Then the situation reduces to that in the previous paragraph.

The perturbation $a \to a'$ of the material input to a metabolic process is not necessarily realized as infection; it may very well be a simple 'change in environment' (e.g. an external stimulus or an internal regulatory trigger). Then the possible reactions of living system H to this perturbation may be interpreted as the stability of H with respect to (external or internal) environmental change. Indeed, various components in H may be dynamical systems with different time scales. Delays in component 'clocks' relative to one another then serve to model, for example, ageing and other dynamical pathophysiologies (*cf.* Section VII.A of

Louie [1985]). The response of H need not be reactive, and may be, rather, anticipatory, when components have internal clocks that run faster than real time. Interacting entailment networks may, in other words, be reformulated as anticipatory systems, but that is the subject for another exposition (cf. *ML*: Chapter 10 and Louie [2012]).

12.5 Alternate Material Entailments   There are three material causes, a, b, and f, in the simplest (M,R)-system H :

(8)

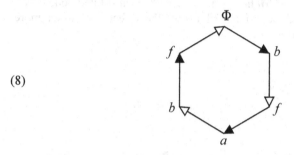

(Indeed, the material causes of the other two classes of (M,R)-systems are also members of the same subset $\{a,b,f\}$ of $\upsilon(H)$.) As for the case of $g \vdash a$ considered in the previous section, similar conclusions may be drawn on the other two material entailments $g \vdash b$ and $g \vdash f$, with interaction diagrams

(9)

(10)

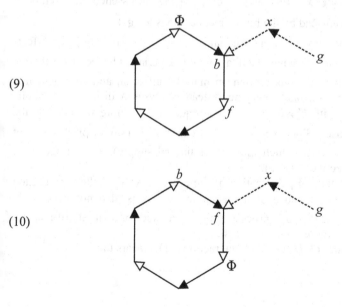

But for these latter two cases, since the g-entailed material causes b and f are also efficient causes in H (with their dual roles $b: f \mapsto \Phi$ and $f: a \mapsto b$, whence $b, f \in \upsilon(H) \cap \kappa(H)$), the effect of this interaction of S on H has further consequences. (Indeed, b and f are final causes in H as well; *cf.* Section 7.6. But it is their additional role as efficient cause that is important here.) This material-cum-efficient cause entailment is a model of prion infection, on which I shall further elaborate later in this chapter.

12.6 Functional Entailment of Metabolism   When the substituted node of H is an efficient cause instead of a material cause, the effects are often more consequential. In the interaction

(11)

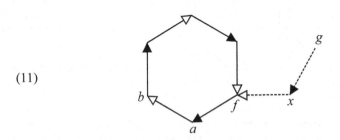

through the action of the process g, S displaces the original enzyme f in H with an antigen $f' = g(x)$; this says $f' \in F_{S \to H}(g)$. In essence, the metabolic entailment $f \vdash$ is replaced by the hierarchical composite $g \vdash f' \vdash$.

If f' acts the same way as the original $f: a \mapsto b$, then the antigenic effects are limited to that of f' being itself a foreign material, whence the situation reduces to that of the previous section on material infection and environmental changes. But if the surrogate enzyme instead catalyses a different metabolic process $f': a \mapsto b'$ with $b' \neq b$, then the consequences are more serious than the 'new metabolite' case. Since a new efficient cause f' is now in place (and not just a new material cause which may be of limited supply), H is infected to produce its own infective material.

Similar to the material perturbation of the previous section, the perturbation $f \to f'$ of the efficient cause of a metabolic process is also not necessarily realized as infection. Metabolic processes may be altered for a variety of reasons, and this is precisely where *repair* enters in.

With all the nodes of H labelled, interaction (11) appears thus:

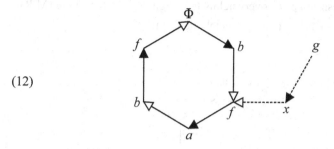

(12)

From the entailment paths, one sees that if the original enzyme f in H has been changed (by the process g of S) into f', the repair $\Phi : b \mapsto f$ may be made, hence replenishing the original f.

12.7 Stability   But note that the material cause (input) of the repair process Φ is b, for this simplest (M,R)-system H. (For a general (M,R)-network, a repair component receives at least one input from the outputs of the metabolism components of the network, that is, the final cause of some $\vdash b$ in the network.) When

(13) $g \vdash f'$ and $f' \vdash b'$,

the input to Φ may be changed accordingly (depending from which $\vdash b$ the repair process Φ draws its input). In this case, the 'recovery' of f (or the stability of H) depends on what Φ entails from the input b', that is, on the value of $\Phi(b') \in \mathrm{Imm}_H(\Phi)$ (where $\mathrm{Imm}_H : \kappa(H) \multimap \kappa(H)$ is the imminence mapping of H; $cf.$ Section 9.1).

Over 50 years ago, Robert Rosen, the creator (or discoverer) of (M,R)-systems, considered the same variational problem [Rosen 1961], albeit from an alterations-in-environment viewpoint rather than my interactions viewpoint. The imminence mapping Imm_H (i.e. the entailment structure of the system H) determines the possible values of $\Phi(b') \in \mathrm{Imm}_H(\Phi)$, which are the following:

i. $\Phi(b') = f \ \left(= \Phi(b)\right)$

ii. $\Phi(b') = f'$

iii. $\Phi(b') = f'' \neq f$ or f'

In the first case, the perturbation $b \to b'$ does not change the output of Φ, so the metabolism processor f is immediately repaired.   In the second case, the replacement metabolism map f' overwhelms the original f, whence the (M.R)-system H changes into H' with a new metabolic form:

(14)

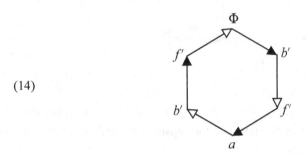

(The reader may have come to the conclusion that in diagram (14), in the entailment of repair Φ, if both perturbations $f \rightarrow f'$ and $b \rightarrow b'$ take hold, then the replication map could have changed from $b : f \mapsto \Phi$ into $b' : f' \mapsto \Phi'$, entailing a new repair map Φ'. I shall explore this scenario in the next section.) In the third case, when yet-another new $\Phi(b') = f''$ is entailed, then the 'stability' of H depends on whether, through iterations of the processes in the (M,R)-network, the sequence of new metabolism maps generated in functional entailment $\vdash f$ is finite or infinite. The metabolic pathways may settle on new stable forms, become periodic, or diverge.

Entailment of Repair and Replication

12.8 Metabolic Entailment of Repair   The   change   in   the   metabolic environment $b \rightarrow b'$ can result, as we saw in the two previous sections, from two different changes in a metabolic process: the perturbation $a \rightarrow a'$ of the material cause or the perturbation $f \rightarrow f'$ of the efficient cause. Since the replication map in an (M,R)-system is $b \vdash \Phi$, one may ask whether $b \rightarrow b'$ would entail $\Phi \rightarrow \Phi'$, that is, whether metabolic interactions or environmental alterations may produce changes in repair mappings that are the genetic processes. An affirmative will have implications on many biological issues, for example, the inheritance of acquired characteristics.

To answer this question, I shall need to consider the nature of the replication map $b \vdash \Phi$ in some detail. The ' b ' appearing herein is a shorthand, a representation of the general replication map β. A replication map must have as its codomain the hom-set $H\big(B, H(A, B)\big)$ to which repair mappings Φ belong, whence it must be of the form

(15) $$\beta : Y \rightarrow H\left(B, H\left(A, B\right)\right),$$

with the domain Y a set already existing in the (M,R)-network. Various choices for Y model different modes of entailment of replication in (M,R)-systems (*cf. ML*: Chapter 12).

For the replication map β of the simplest (M,R)-system (which belongs to the first class of entailment of replication), $\beta \cong b$ after the isomorphic identification

(16) $$b \cong \hat{b}^{-1}$$

is made between $b \in B$ and the *inverse evaluation map*

(17) $$\beta = \hat{b}^{-1} \in H\left(H\left(A, B\right), H\left(B, H\left(A, B\right)\right)\right)$$

(*ML*: 11.16). This mode of replication hinges on the existence of the inverse of the evaluation map $\hat{b}$ (under a condition that is the algebraic formulation of the *one-gene-one-enzyme hypothesis*). When $b \rightarrow b'$, the evaluation map changes correspondingly from $\hat{b}$ to $\hat{b}'$. One readily verifies that if the inverse $\hat{b}'^{-1}$ exists, then

(18) $$\hat{b}'^{-1}\left(\Phi\left(b'\right)\right) = \Phi,$$

and the repair component Φ is exactly replicated, so the repair map is repaired, and the system 'recovers'. The structures of the hom-sets $H\left(H\left(A, B\right),$ $H\left(B, H\left(A, B\right)\right)\right)$ determine the existence of inverse mappings. A category-theoretic argument shows that, for appropriate hom-sets, if $\hat{b}^{-1}$ exists for some b, then it exists for all b. The argument is predicate on the hom-sets being 'not too large', and the existence of the inverse evaluation map depends on the functional stability (*cf.* Section 9.7) of the processes of the (M,R)-system. In particular, the hom-sets $H\left(X, Y\right)$ in the category of models containing the (M,R)-system must at the very least be a proper subset of the set $Y^X = \mathbf{Set}(X, Y)$ of all mappings from set X to set Y —evaluation maps in the category $\mathbf{Set}$ are not in general invertible.

Note that the situation when $b \rightarrow b'$, entailing $\Phi\left(b\right) \rightarrow \Phi\left(b'\right)$ accordingly, is somewhat different from a direct perturbation of $f \rightarrow f'$ in $H\left(A, B\right)$. While the natural recovery of Φ from $b \rightarrow b'$ is an algebraic consequence of the relational organization of an (M,R)-system, additional assumptions are required to make a genetic component Φ invariant to a direct $f \to f'$. This is, however, not to say that mappings somehow have 'memories' of their entailments; it has to do, rather, with the limiting effect of ranges. For a mapping to be functionally entailed implies that it must belong to the range of its entailer; this is yet-another illustration of the proper containment relation $H(X,Y) \subset \mathbf{Set}(X,Y)$, and containment is by definition restrictive. From a set-valued mapping point of view, the entailed perturbation $\Phi(b) \to \Phi(b')$ depends on the behaviour of the specific mapping $\mathrm{Imm}_H : \kappa(H) \multimap \kappa(H)$, the imminence mapping of H, while a direct perturbation of $f \to f'$ may be any general mapping in $\mathbf{Rel}(\kappa(H), \kappa(H))$.

In summary, for organisms that are realizations of the (most common) first class of (M,R)-systems, acquired metabolic and physiological changes mostly, under stringent but not prohibitive conditions, do not lead to genetic alterations and are hence not inherited.

12.9 Alternate Entailments of Replication While the first class of entailment of replication is functionally stable with respect to metabolic entailment of genetic change, the second class is different. The replication map here is

$$(19) \qquad \beta = \gamma_B \in H\left(B, H\left(B, H\left(A,B\right)\right)\right)$$

(where γ_B is a Hilbert space conjugate isomorphism; cf. ML: 12.7), with a one-to-one correspondence $b \leftrightarrow \Phi$. Indeed, the interaction $\langle b, \Phi \rangle$ may be realized as *enzyme specificity* in enzyme–substrate recognition processes (cf. ML: 12.9). (Note, however, that the bijection $b \leftrightarrow \Phi$ is a mathematical result; biology is full of exceptions, and there is evidence that enzyme specificity is not absolute. When it comes to biology, all absolute statements are false, including this one.) A perturbation $b \to b'$ would entail correspondingly $\Phi \to \Phi'$. So in this case, a perturbation of a metabolism component (just as a direct perturbation of the metabolic process $f \to f'$) leads to a change in a genetic process; the effect of S on H (i.e. the new environment or the interaction) is therefore *mutagenic*.

For the third class of entailment of replication,

$$(20) \qquad \beta = \pi_S \circ (\,\cdot\,)^{-1} \in H\left(A, H\left(B, H\left(A,B\right)\right)\right)$$

(where $\beta : a \mapsto \left[a^{-1}\right]_S = \Phi$ is a mapping of similarity classes and models protein biochemistry; cf. ML: 12.12). A perturbation $a \to a'$, therefore, may or may not alter the output Φ, depending on whether $\left[a'^{-1}\right]_S$ and $\left[a^{-1}\right]_S$ are the same similarity class. The entailment of replication in (M,R)-systems in this mode models alterations in protein chemistry and may have epigenetic effects (e.g.

nucleic acid methylation and histone acetylation) that affect gene expressions or may even directly or cumulatively cause mutations.

I have deliberately not invoked the term 'Lamarckism' in the foregoing discussion, its use/misuse being a whole separate topic. Let me just say that Jean-Baptiste Lamarck was not a Lamarckian (just as Isaac Newton was not a Newtonian and Charles Darwin was not a Darwinian).

Natural Selection

12.10 Epigenetics   We have heretofore seen that a metabolic alteration, be it $a \to a'$, $b \to b'$, or $f \to f'$, may or may not engender a genetic change $\Phi \to \Phi'$ in the (M,R)-system under consideration.   The pathophysiological response depends on the topology of the (M,R)-system, on the specific entailment pathways that culminate in $\vdash \Phi$.   Both negative and positive responses are, indeed, observed in living systems.   Many interactions and environmental changes have harmful but recoverable effects.   Others, although not themselves genetic changes, either have epigenetic effects that affect genetic expressions without mutations or be ultimately mutagenic, leading to, for example, cancer and heritable diseases.

The adverb 'ultimately' is important in the previous sentence.   This is because mutations may be symptoms and not immediate effects, contrary to what one may often believe.   For example, it is an implicit assumption that ionizing radiation directly damages the DNA molecule.   But perhaps the effect is not so reductionistically molecular: what is damaged by radiation may be a cellular process, and DNA structural damage was simply a symptom of this functional damage, a link further down the causal chain.   (Recall that 'function dictates structure' is the *modus operandi* of relational biology, as opposed to the 'structure implies function' of molecular biology.)   With organisms as realizations of (M,R)-systems, one may readily recognize that what is directly damaged may have been a metabolic process that happens to have epigenetic effects; mutation is manifested when the epigenetic perturbation causes a genetic alteration that the replication map β cannot repair.

12.11 Mutagenesis   Some mutagenic environments may, of course, affect cells by directly causing mutations.   Stated otherwise, the entailment of mutation $\Phi \to \Phi'$ can also be explicit. Let S contain a process g that affects an internal process f of H on an even higher hierarchical level still, as the genetic interaction

(21)

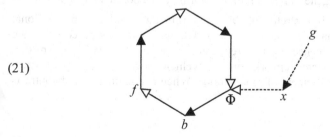

in which g entails a final cause that replaces the original gene Φ with the rebel $\Phi' = g(x)$, that is, $\Phi' \in F_{S \to H}(g)$. Now, in the host H, the new genetic expression $\Phi' : b' \mapsto f'$ replaces the old $\Phi : b \mapsto f$ and produces new copies of the antigenic enzyme f' for further metabolic devastation.

Recall (Section 12.2) that the two entailment networks H and S are not necessarily separate: S may simply be a subsystem of H, partitioned for the study of its perturbation effects. In this formulation, the mutation $\Phi \to \Phi'$ may simply be an 'internal' replication error, with $g \in \kappa(H)$, $\Phi' \in \mathrm{Imm}_H(g)$, and $f' \in \mathrm{Imm}_H^2(g)$. The processes of metabolism–repair–replication are ordained *in principio* to make small mistakes. The capacity to err is, in fact, the real marvel of evolution; through mutational blunders, progress and improvements are made. The Latin root for 'error', the driving force of evolution, is *erratio*, which means roving, wandering about looking for something, quest.

12.12 Relational Evolution Mutagenesis, the process of change of the genetic information of an organism, is the driving force of evolution. Just as the perturbations on other hierarchical levels, the mutational 'error' $\Phi \to \Phi'$ is not necessarily realized as a harmful change. What the process g entails may be interpreted as a *gene exchange* between S and H and may, for example, be a relational model of homologous recombination in meiosis. Indeed, gene exchange is the very definition of *sex* in its most general biological terms. It has been postulated that sex possibly evolved as a mechanism to facilitate the repair of damaged genes. Viruses are also an important means of horizontal gene transfer, thus increasing genetic diversity. In the gene exchange that g functionally entails, H receives an 'intact' gene Φ' from S to replace its own 'broken' gene Φ. In other words, the merits of the 'normal' Φ and the 'abnormal' Φ' may be reversed, and the irregular Φ' instituted by the process g of S on H may in fact be beneficial to H, either as repair (i.e. the 'R' in (M,R)-systems) or as a heritable characteristic that, when manifested, improves the competitive fitness of the organism H in its environment.

In its most general form, natural selection is the choice among alternative heritable characteristics, that is, the competitive fitness of 'replicators'. The process selects replicators that are more proficient in their replications, the proficiency implying in particular their survival advantage over their competitors. We have now learned how this 'efficient cause of the efficient cause of evolution' may be formulated as hierarchical entailment in interacting (M,R)-systems; the replicator $\Phi \in F_{S \to H}(g)$ is a 'gene' in its relational form (as opposed to the archetypal molecular form that is a stretch of DNA). From the standpoint of relational biology, the unit of natural selection is the functionally entailing process (i.e. the mapping and its imminence) on the 'replicator' level and the (M,R)-system (i.e. clef functional entailment network) on the 'vehicle' level (i.e. the hierarchical level of organisms, containers of replicators). When one compares and contrasts this formalism with the emergent steps in relational genesis (Section 8.28), one sees clearly that the processes of the evolution of life and the origin of life are different in kind.

Genesis, as previously explained, is an event that needs to happen only once; contrariwise, evolution is an ongoing serial process. Biopoiesis is a probabilistic consequence of the developing and increasing complexity of entailment networks, and natural selection is a deterministic consequence of (probabilistic mutations in) interacting entailment networks. In biogenesis, the improbability drive from nonlife to life is overcome once (or feasibly more times), entailed through repetition and strong laws of almost sure convergence. Events of adaptive fit, on the other hand, act cumulatively. The end product of the evolutionary accumulation, after a large number of mutational events, may appear very improbable. But each singular event is only slightly improbable (if not indeed simply apparently so) and occurs randomly as a matter of routine in the natural entailment inherent in living systems. In very terse predicate terms, genesis is $\exists$ and evolution is $\forall$. Formulated with respect to imminence mappings, genesis is their becoming and evolution is their being: from nonlife to life, the imminence mapping Imm_N requires opportune poiesis to become clef, but once clef, the continual natural selection among the Imm_N s of (M,R)-systems N, through interactions of mutual functional entailment $F_{.\to.}$, is a natural consequence of their being clef.

Origins

12.13 Synthetic Life For fundamental logical reasons, the kind of 'synthetic biology' in which many people are involved—the mechanistic, algorithmic, and by-part attempts at the fabrication of life—will not work. These ventures invariably amount to the transformation of one life form to another (although the latter may have never per se existed before). If their definition of 'synthetic biology' is modified to be 'assembly of new life forms', then such activities are simply the carrying-on of a time-honoured tradition. There is nothing new about the modification of existing organisms; people have been doing that for millennia, at least since the dawn of agriculture.

In the twenty-first century, we simply have more sophisticated tools. Examples of genetic engineering tools include the now-commonplace partial modification of the genome by multiple insertions, substitutions, or deletions, or even the synthesis of the entire genome but, alas, all within the confines of the 'vehicle' of an already living cell. But 'remodelling' is not the same thing as 'new construction'. Results of these molecular techniques are still extremely remote from the true *telos* of synthetic biology, that (by definition) of 'life from nonlife', *de novo* ('from scratch'). That has not, unfortunately, stopped overreaching and artificial claims of the hyperbolic 'accomplishment' of having created life.

In strictly reductionistic physicochemical terms, biology-from-molecules is a highly improbable stochastic event that is whim-prone. There is no way to tell which exact molecules were involved in the critical step in molecular evolution on this planet in the passage from nonlife to life. Was it from amino acids to proteins? Was it from nucleotides to nucleic acids? Or perhaps it was among some archaeobiological molecules that had long since disappeared from the biosphere. Even if one had precisely the correct molecules, how about the very environment that allowed the chemical reaction to take place? The critical step was more likely a process that had nothing to do with trivial polymerisation but, rather, something that has no direct correspondence in molecular terms and is more than what we could have "dreamt of in our philosophy". The futility of algorithmic synthetic biology is, in any case, not just a matter of molecular improbability in the laboratory. Biochemistry has progressed so far and so fast in the past century that people find it hard to imagine that the process cannot continue *ad infinitum*. The main problem is that the reductionist biology-is-chemistry approach has been so successful in solving biological puzzles that although everyone can recognize that a living system is not just a machine, there is a great reluctance to admit that the two are different in kind and not just in degree.

Many older problems in biology have merely been displaced, but not solved, by the explosive developments of approaches at the molecular level. These problems involve the very core of biological organization and development, for example, homeostasis, ontogenesis, phylogenesis, consciousness, and, of course, the ultimate biological question: "What is life?" One must not lose sight of these problems, and one must seek out alternative (other than reductionistic molecular) approaches to address them.

As I (and we in the Rashevsky–Rosen school of relational biology) have repeatedly emphasized, biology is a subject concerned with organization of relations. A living system is a material system, so its study shares the material cause with physics and chemistry. But physicochemical theories are only surrogates of biological theories, because the manners in which the shared matter is organized are fundamentally different. Hence, the behaviours of the realizations of these mechanistic surrogates are different from those of organisms. This in-kind difference is the impermeable dichotomy between predicativity and impredicativity.

In short, 'ontological arguments' of synthetic life, algorithmically from the material cause of molecules, are doomed in both degree and in kind.

12.14 Trial and Error   Even when genesis is formulated in relational terms, the synthesis of life from nonlife is still a probabilistic 'almost sure convergence' event. I leave it as an exercise to the reader to calculate the probability of typing monkeys producing the phrase *Functional Entailment and Imminence in Relational Biology* and to use the result to verify the Second Borel-Cantelli Lemma (*cf.* Section 8.1). Now consider that the origin of life, relational (*cf.* the three emergent steps of functional entailment and imminence; Section 8.28) or otherwise, is a great deal more improbable than that. The same almost sure convergence principle applies, however, and the event only has to happen once, in an almost infinite sequence of repeat-until-it-works. The fabrication of life from nonlife is probable (and therefore possible), but it cannot be achieved by rote, algorithmic means.

Sidney Harris, the science cartoonist, has drawn a comic strip that illustrates the subject perfectly (well, almost perfectly, if only, alas, it were not so reductionistically molecular...):

Reprinted with permission from ScienceCartoonsPlus.com

Stochastic implications are like that: sure, one knows that in a suitably equipped category, the emergence of (M,R)-systems from large functional entailment networks is almost inevitable, but one wouldn't wait around to observe it.

12.15 Anthropic Principle   But biogenesis did happen, at least once. We are here to formulate the theories, aren't we?   This self-referential argument is, incidentally, known as the 'anthropic principle' [Barrow & Tipler 1986]. It may be said that, in philosophy, 'a universe forever empty of life' is inconceivable, a contradiction. Quite simply, as John A. Wheeler wrote in the Foreword to Barrow and Tipler [1986], 'It has no sense to talk about a universe unless there is somebody there to talk about it.'

The issue is not as flippant as it may appear. Indeed, the two axioms of Natural Law (*cf. ML*: 4.7) allow, respectively, science and scientists to exist. The modelling relation argues a modeller.

Werner Heisenberg wrote the essay *Das Naturbild der heutigen Physik* in 1954. It was first published in the 1956 book *Die Künste im Technischen Zeitalter*. The English translation by O. T. Benfey, *The Representation of Nature in*

Contemporary Physics, appeared in *Daedalus* (vol.87 no.3, *pp*.95–108) in 1958 and was reprinted in several collections:

> ...we are finally led to believe that the laws of nature which we formulate mathematically in quantum theory deal no longer with the particles themselves but with our knowledge of the elementary particles. The question whether these particles exist in space and time "in themselves" can thus no longer be posed in this form. We can only talk about the processes that occur when, through the interaction of the particle with some other physical system such as a measuring instrument, the behavior of the particle is to be disclosed. The conception of the objective reality of the elementary particles has thus evaporated in a curious way, not into the fog of some new, obscure, or not yet understood reality concept, but into the transparent clarity of a mathematics that represents no longer the behavior of the elementary particles but rather our knowledge of this behavior. The atomic physicist has had to come to terms with the fact that his science is only a link in the endless chain of discussions of man with nature, but that it cannot simply talk of nature "as such". Natural science always presupposes man, and we must become aware of the fact that, as Bohr has expressed it, we are not only spectators but also always participants on the stage of life.

While Heisenberg was speaking to atomic physicists on quantum theory, the lesson may well be learned by biologists on life itself. The Heisenberg uncertainty principle

$$(22) \qquad\qquad \sigma_x \sigma_p \geq \frac{\hbar}{2}$$

is a probabilistic statement (σ being the standard deviation) and is a fundamental property of nature. (It has nothing to do with the 'observer effect'.) The uncertainty inherent in our being "participants on the stage of life" renders nature itself and our interactions with nature, both, impredicative.

Impredicative (nonalgorithmic) consequences of human-in-the-loop coupling in human–machine interface (HMI) are explored in a special topical issue of the *Journal of Integrative Neuroscience* (volume 4 number 4, 2005; [Kercel 2005]): "brains must go beyond computing". Generalized quantum theory (GQT), the relaxation of quantum theory to encompass macroscopic systems and the exploration of the transdisciplinary applicability of quantum-theoretic concepts, is the theme of a special topical issue of *Axiomathes* (volume 21 number 2, 2011 [Walach & von Stillfried, 2011]).

13
Relational Virology

Il est peut-être plus juste de considérer les virus comme des processus d'auto-destruction de la vie empruntant un formalisme vital, des chréodes suicidaires du métabolisme cellulaire.

[Perhaps it is better to consider viruses as self-destructive processes of life taking on a living formalism, suicidal chreods of cellular metabolism.]

> — René Thom (1972)
> *Stabilité structurelle et morphogénèse :*
> *Essai d'une théorie générale des modèles*
> Chapitre 12
> Les grandes problèmes de la biologie

Pathogeny

13.1 Suicidal Chreod A chreod (from Greek χρεών necessity and οδός pathway; literally 'necessary path') is a steady trajectory that acts as an attractor for neighbouring trajectories. The term was coined by C. H. Waddington in connection with homeorhesis (when dynamical systems return to such steady trajectories) in morphogenetic fields in embryology and was later adopted by René Thom in general topological terms in his catastrophe theory of structural stability.

We have already encountered viruses in connection with symbiosis (in Sections 11.4 and 11.11). In this chapter I shall develop further a theory of the relational biology of these obligate intracellular parasites. When a virus infects, it initiates its self-destruction sequence, takes on the living (i.e. (M,R)-) formalism of a cell, possesses the cellular metabolism for its own replication, and leads the cell down a suicidal chreod towards the release of its progeny. In short, the relational scheme is exactly as Thom considered above.

A.H. Louie, *The Reflection of Life*, IFSR International Series on Systems Science and Engineering 29, DOI 10.1007/978-1-4614-6928-5_13,
© Springer Science+Business Media New York 2013

13.2 Viral Infection   Let H and S continue to represent two interacting entailment networks. The genetic interaction

(1)

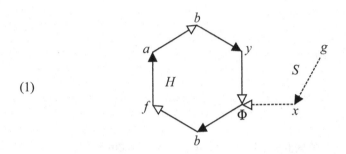

between H and S merges the two entailment networks S and H into their *join* entailment network $H \vee S$ (*cf. ML*: 7.28). The original functional entailment

(2) $$\Phi \vdash f$$

in $\kappa(H)$ is replaced by the hierarchical composition

(3) $$g \vdash \Phi' \vdash f'$$

in $\kappa(H \vee S)$. This explicit change $\Phi \to \Phi'$ in genetic expression in the (M,R)-system H is how a virus S infects, imposing its genes $\Phi' \in F_{S \to H}(g)$ on the host H and then subsequently (and hierarchically) using the host's metabolic–repair–replication processes to produce more copies of Φ' (*cf.* Section 10.6 for a review of the *inter-network imminence* of S on H, $F_{S \to H} : \kappa(S) \multimap \kappa(H)$). In essence, the infected H has its hierarchical cycle $\langle f, \Phi, b \rangle \subset \kappa(H)$ modified to $\langle f', \Phi', b' \rangle \subset \kappa(H \vee S)$.

13.3 Replicator   While viruses are not themselves alive, when inside the host cell they become alive. It is only after the initiation of infection, whereupon a virus appropriates the host's 'metabolism' component (indeed all of the host's metabolic–repair–replication processes) for the replication and expression of its viral genome (entailing subsequent release of mature virions from the infected cell) that the virus becomes 'alive', in the sense that its 'repair' (i.e. genetic) component is now part of a clef (M,R)-system. Teleologically, the sole objective of a virus is to replicate the genetic information contained in this repair component, i.e. to express the functional entailment therein, to manifest its imminence. But then, every organism may arguably be said to have the same objective;

'replication'—or at least the propagation of the genome contained therein—is the ultimate *telos* of life. Succinctly, the difference between the replications of viruses and living systems is that of voice, in the sense of passive versus active: viruses are replicated; cells replicate themselves. The intimate symbiosis with living systems earns the virus the honorific title of *biological entity*.

Viruses are found throughout the biosphere and are, indeed, the most abundant kind of biological entity. They infect all other life forms, bacteria, fungi, plants, and animals. Each type of virus, however, can only infect a limited number of species (its *host range*, which may be broad or narrow), and some are species-specific (having a very narrow range of one). It may (although rarely) happen that a virus mutates and crosses species to infect a previously unaffected host species.

In addition to being efficient replicators, viruses are the original 'synthetic biology' practitioners: in viral infection, the virus S modifies the existing (M,R)-system H to the new host+virus (M,R)-system $H \vee S$. But note that this transition $S \to H \vee S$ is not synthetic biology in the life-from-nonlife sense— even viruses cannot overcome that algorithmic impossibility. The living system $H \vee S$ is not assembled *de novo*; the inanimate S simply 'remodels' the existing dwelling H for its purposes.

Virus Components

13.4 Empty Imminence   Recall (Definition 7.10) that an (M,R)-network N is the entailment network of a finite collection of pairs of metabolism and repair components $\{(M_i, R_i) : i \in I\}$, where the metabolism and repair components are, respectively, the formal systems $M_i = \langle A_i, H(A_i, B_i) \rangle$ and $R_i = \langle Y_i, H(Y_i, H(A_i, B_i)) \rangle$. The components M_i and R_i may be considered *subnetworks* of the (M,R)-network N and are therefore entailment networks in their own right. In particular, the collections of metabolism and repair maps of the components are subsets of the processes of N (*cf.* Section 7.13):

$$H(A_i, B_i) \subset \kappa(M_i) \subset \kappa(N) \text{ and } H(Y_i, H(A_i, B_i)) \subset \kappa(R_i) \subset \kappa(N).$$

A repair map $q \in \kappa(R_i)$ is 'that which functionally entails', and a metabolism map p 'that is functionally entailed' in $q \vdash p$ must be from the corresponding metabolism component, $p \in \kappa(M_i)$. Symbolically, this says

(4) $$\forall q \in \kappa(R_i) \; \exists p \in \kappa(M_i): \; p \in F_{R_i \to M_i}(q)$$

(where $F_{R_i \to M_i} : \kappa(R_i) \multimap \kappa(M_i)$ is the inter-network imminence of R_i on M_i; *cf.* Section 10.6). This 'repair $\vdash$ metabolism' entailment may be tersely represented as

$$(5) \qquad\qquad R_i \vdash M_i.$$

A fortiori, to be an (M,R)-network implies that

$$(6) \qquad\qquad \forall\, q \in \kappa(R_i)\ \ F_{R_i \to M_i}(q) \neq \varnothing.$$

In contrast, a virus S in isolation consists only of a repair component R_i (without its corresponding metabolism component M_i), i.e. a virus has 'empty imminence' $F_{R_i \to M_i}(q) = \varnothing$. More precisely, this empty imminence is formulated in this way: a virus is $S = R_i$ with $\kappa(M_i) \cap \kappa(S) = \varnothing$, whence $F_{R_i \to M_i}(q) = \kappa(M_i) \cap \mathrm{ran}(q) = \varnothing$. Thus, I make the following relational–biological

13.5 Definition A *virus* is an entailment network containing a repair component but not the corresponding metabolism component.

Equivalently, a virus S may be considered a networked collection of repair components without the corresponding metabolism components. The crucial characterization is that the 'repair effect' (generalized functional entailment; *cf.* Section 10.6) of each process $q \in \kappa(S)$ is empty:

$$(7) \qquad\qquad F_{S \to S}(q) = \mathrm{Imm}_S(q) = \varnothing.$$

The 'metabolism effect' $M_{S \to S}(q)$ (generalized material entailment; *cf.* Section 10.5) of a process $q \in \kappa(S)$, on the other hand, may be nonempty (comprising the protective and attachment devices in the virus architecture).

The processes of the virus S awake from their dormancy once S has invaded the host H. The parasite effects the change $f \to f'$ in the host's metabolism by replacing the original imminence

$$(8) \qquad\qquad f \in \mathrm{Imm}_H(\Phi)$$

in the (M,R)-system H with the viral imminence

$$(9) \qquad\qquad f' \in \mathrm{Imm}^2_{H \vee S}(g)$$

in the new host+virus (M,R)-system $H \vee S$ (*cf.* the functional entailments (2) and (3) above, and Section 9.9 for the iterated imminence Imm^2). Note that the imminence mapping Imm_S of the singular repair component that is the virus S already contains the seed of the infective indirect imminence

(10) $\text{Imm}^2_{\cdot \vee S} = F_{S \to \cdot} \circ \text{Imm}_S$.

13.6 Virus Architecture   *Viroids* indeed comprise nothing but an isolated repair component. *Satellites* are obligate viral parasites (fleas have fleas *ad infinitum* ...!) that are dependent on the presence of another virus for replication. Both of these infective agents are materially realized as small RNA molecules and may be considered *minimal viruses* under our relational definition.

Most virus particles (*virions*), however, contain more than just the genome. They have a protective protein coat (*capsid*) that protects the genetic component, and some have an envelope of lipids that surrounds the protein coat when the viruses are outside a cell. In addition to its physicochemical protection role, the outer surface of viruses also facilitates their attachment to the cells on which they are parasitic. The protein capsid may even be responsible for more than the recognition and the initial interaction with the host cell surface: in some cases, the viral genome undergoes extensive modifications before insertion into the host, and there is evidence that these processes are aided by the capsid proteins.

In a general formulation, one may consider an entailment network

(11) $N = \left\{ M_i : i \in I \right\} \cup \left\{ R_j : j \in J \right\}$

as the union of two separate collections, each of a different kind of formal systems, $M_i = \left\langle A_i, H\left(A_i, B_i \right) \right\rangle$ and $R_j = \left\langle Y_j, H\left(Y_j, H\left(A_j, B_j \right) \right) \right\rangle$. The M_is are general 'metabolism' components that are *material entailers*, and the R_js are general 'repair' components that are *functional entailers*. An (M,R)-network is an entailment network (11) specialized with its finite collections of metabolism and repair components paired, $I = J$, whence represented as $N = \left\{ \left(M_i, R_i \right) : i \in I \right\}$ and with the mandated $R_i \vdash M_i$ for each $i \in I$. A virus N is (11) specialized with mutually exclusive index sets, $I \cap J = \varnothing$, so that if $R_j \in N$, then $M_j \notin N$, but it is entirely feasible to have extraneous functional entailments $M_i \vdash R_j$ for $i \neq j$ in $\kappa(N)$.

13.7 Clef Closure   The transition $S \to H \vee S$ from a virus S, as its infects its host the (M,R)-system H , to the new host+virus (M,R)-system $H \vee S$ is somewhat more complex than the genetic interaction (1) of two entailment networks S and H . The prerequisite is, of course, that the networks do interact, which is the statement

(12) $$F_{S \to H}(g) = \kappa(H) \cap \operatorname{ran}(g) \neq \varnothing,$$

when formulated in the inter-network imminence of S on H. One may observe that although the entailment network (1) is complex (since it contains a hierarchical cycle; cf. ML: 9.2), it is not clef (since the mapping g is not functionally entailed; cf. ML: 6.26).

The infective process $g \in \kappa(S)$ of the virus S is such that $g \vdash \Phi'$, with Φ' replacing an existing genetic process $\Phi \in H\left(B_j, H\left(A_j, B_j\right)\right) \subset \kappa\left(R_j\right) \subset \kappa(H)$ of the host cell H. One may, without loss of generality, let the virus S contain the repair component $R'_j = \left\langle X, H\left(X, H\left(A_j, B_j\right)\right)\right\rangle \subset S$ but $M_j = \left\langle A_j, H\left(A_j, B_j\right)\right\rangle$ $\not\subset S$. For a host H to be susceptible to infection by S, it must already contain in its own (M,R)-system the pair of metabolism–repair components $\left(M_j, R_j\right) = \left(\left\langle A_j, H\left(A_j, B_j\right)\right\rangle, \left\langle Y_j, H\left(Y_j, H\left(A_j, B_j\right)\right)\right\rangle\right)$ with

(13) $$Y_j \cap X \neq \varnothing.$$

This nonempty intersection serves as a necessary and sufficient condition for the infection and is the relational explanation of the host range of a virus.

The infective process $g \in \kappa(S)$ does not have to be a process in R'_j itself (i.e. it does not have to be solely a 'replication process'). Remember that in (M,R)-networks there is no one-to-one correspondence between functional processes and the structures that realize them (cf. Section 7.7 & ML: 11.34). All that is required for g is for the inter-network imminence of S on H, $\Phi' \in F_{S \to H}(g)$, to potentially proceed. Indeed, the process '$g \vdash$' may represent all the initiation steps in the virus–cell interaction, with genetic perturbation '$\vdash \Phi'$' as its final, final cause.

Since the host $H = \left\{\left(M_i, R_i\right) : i \in I\right\}$ itself is clef, its repair component $R_j = \left\langle Y_j, H\left(Y_j, H\left(A_j, B_j\right)\right)\right\rangle$ is 'replicated', i.e. entailed within H. The condition (13) then implies that the viral repair component $R'_j = \left\langle X, H\left(X, H\left(A_j, B_j\right)\right)\right\rangle$ is also replicated. In particular, one has '$\vdash g$' for those infective processes $g \in \kappa(S)$, for which $F_{S \to H}(g) \neq \varnothing$. The expression of the viral genome $\Phi' \in F_{S \to H}(g)$, thus, entails reproduction of the virus, $g \in \operatorname{Imm}_{H \vee S}^n(\Phi')$ (with the path length n depending on the exact hierarchy of functional entailments to arrive at '$\vdash g$'). In short,

(14) $$H \vee S \vdash S,$$

whence the host+virus (M,R)-network $H \vee S$ achieves 'clef closure' and may be considered an (M,R)-system, i.e. a living system itself.

But of course, with the subsequent release of mature virions from the infected cell $H \vee S$, this particular host+virus (M,R)-system will no longer exist. The entailment $H \vee S \vdash S$ rings the death knell of $H \vee S$, and the iteration $S \rightarrow H \vee S$ carries on.

13.8 Origin of Viruses   Viruses probably existed since the first living cells appeared, but the lack of fossilized evidence means that the origins of viruses can only be postulated. Common hypotheses are that viruses originated:

i.   In a precellular world from macromolecules ['virus-first hypothesis']

ii. By devolution from parasitic cells ['reduction hypothesis']

iii. From cellular genetic fragments that are breakaways ['escape hypothesis']

All three hypotheses have specific deficiencies and are reductionistically molecular (although 'molecular virology' is not the oxymoron that 'molecular biology' is).

From the standpoint of relational biology, we may forego the materialistic origins and simply consider that organisms (i.e. general living systems) and viruses are two different specializations of the entailment network (11). Thus, organisms and viruses arose on separate branches of the *lignum vitae* that realized different emergent features in the relational genesis of mutually entailing efficient causes (*cf.* Section 8.28). The relational taxonomy may be concisely formulated in terms of the imminence mapping $\mathrm{Imm}_N : \kappa(N) \dashv\subset \kappa(N)$. A living system N is clef, an (M,R)-system in which

(15) $$\forall f \in \kappa(N) \quad \mathrm{Imm}_N^{-1}(f) \neq \varnothing$$

(*cf.* Theorem 9.2). A virus N is an obligate parasite, an entailment network of R-without-M, whence

(16) $$\forall f \in \kappa(N) \quad \mathrm{Imm}_N(f) = \varnothing.$$

These conditions (15) and (16), of course, put different restrictions on the imminence mapping Imm_N and the processes in $\kappa(N)$. The in-kind differences and their entailments may provide a feasible explanation on why the available processes in the *biosphere* and the *virosphere* are vastly different.

Infection Modes

13.9   Functional Entailment on Three Hierarchical Levels   Consider the generic interaction diagram

(17)

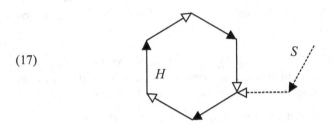

whence the symbiont S contains a process $g \in \kappa(S)$ that functionally entails a process of the host H. When H is the unfolded hierarchical cycle

(18)

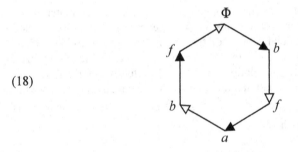

of the simplest (M,R)-system, the process g may entail any one of the three maps $\{f, \Phi, b\} \subset \kappa(H)$. When formulated in terms of the functional entailment set-valued mapping $F_{S \to H} : \kappa(S) \multimap \kappa(H)$, one of these three different modes of infection potentially happens if

(19) $\{f, \Phi, b\} \cap F_{S \to H}(g) \neq \varnothing .$

The functional entailment

(20) $g \vdash f ,$

the connection of S to H at node f, models functional entailment of metabolism (cf. Section 12.6). When the connection is at node Φ, one has the compounded functional entailments

(21) $g \vdash \Phi \vdash f$,

and this mode of interaction between S and H models mutagenesis (cf. Section 12.11) and viral infection (cf. Section 13.2 above). The third possibility is

(22)

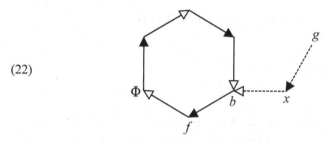

with common node b and entailment

(23) $g \vdash b$.

13.10 Prion Infection   In Section 12.5, I noted that, in the simplest, standard (M,R)-system H, b and f are material causes as well as efficient causes. So when $g \vdash b$ or $g \vdash f$, the material entailment of $b, f \in \upsilon(H)$ and functional entailment of $b, f \in \kappa(H)$ are synodal processes. This is to say, the material perturbation $b \to b'$ and the functional perturbation $\hat{b}^{-1} \to \hat{b}'^{-1}$ are linked and likewise for $f \to f'$ as a (material) protein or as a (functional) enzyme. This subtle linkage of structure and function at opposite nodes in the unfolded (M,R)-system (18) serves to model the process of prion infection.

   A *prion* (by rearrangement from '*pro*teinaceous *in*fectious particle') is an infectious agent composed of protein in a misfolded form. This is in contrast to all the other infectious agents that I have thus far modelled, since they (parasite, bacteria, fungus, and virus) all contain genetic components in their (M,R)-networks. When a prion enters a healthy organism, it induces existing properly folded proteins to convert into the disease-associated misfolded form.

   The *prion hypothesis* postulates that there is a source of *information* within protein molecules that contributes to their biological function and that this information can be passed on to other molecules. In the case of a prion protein, it acts as a template to guide the misfolding into prion form of more molecules of the normally folded form of the same protein (with the prerequisite that the infected organism already contains the latter). The prion hypothesis is not heretical to the central dogma of molecular biology (which states that the information necessary to manufacture proteins is encoded in the nucleotide sequence of nucleic acid and its flow is unidirectional), because it does not claim that proteins replicate. Rather, it claims that there is a source of information within protein molecules that contributes to their biological function and that this information can be passed on to other molecules; the protein molecules are still manufactured according to the instructions contained in nucleic acid.

This non-genetic 'information' within protein molecules is readily explained in our context of relational biology. The isomorphism

$$(24) \qquad\qquad B \cong H\big(H(A,B), H\big(B, H(A,B)\big)\big),$$

i.e. the map

$$(25) \qquad\qquad\qquad b \mapsto \hat{b}^{-1},$$

is the source of this information. We have already seen in Section 12.8 that the material perturbation $b \to b'$ and the entailed functional perturbation $\hat{b}^{-1} \to \hat{b}'^{-1}$ do not affect the stability of the genetic map Φ. Stated otherwise, the map (25) establishes between the two ' b ' nodes in the unfolded (M,R)-system (18) a correspondence that bypasses the node Φ.

The immediacy of this correspondence is more evident if one considers the interaction with the (M,R)-system in the folded multigraph form, when there is in fact just one node b :

(26)

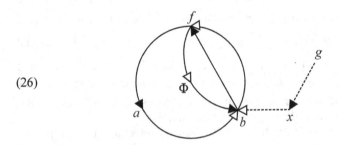

Similarly, the immediate non-genetic correspondence between the material cause f and the efficient cause f is clearly illustrated in

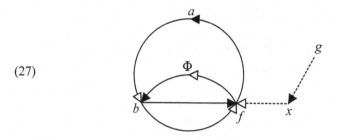

(27)

13.11 Pseudographs   When the (M,R)-network H is in the second or third class of (M,R)-systems (*cf. ML*: Chapter 12), the respective isomorphisms that describe the dual roles of material–efficient causes also serve to model the non-genetic information carried within protein molecules.   The category-theoretic details are presented below, but it is more important to see beyond the mathematical technicalities of the three isomorphic correspondences defined by the $(\div)^{-1}$, $\gamma_B(\cdot)$, and $\pi \circ (\cdot)^{-1}$. One may more meaningfully note that in all three classes of (M,R)-systems, there is a natural way (in both the common-usage and category-theoretic senses) for a material cause to have a dual role and carry information as an efficient cause.

The functional entailment $F_{S \to H}(g)$ when H is in the second class of (M,R)-systems has $b \in F_{S \to H}(g)$ and the relational diagram

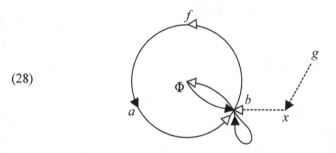

(28)

Here, the non-genetic information within protein molecules is carried by the conjugate isomorphism (*ML*: 12.7)

(29) $$ B \cong H\big(B, H\big(B, H\big(A, B\big)\big)\big), $$

the dual roles of material–efficient causes for b manifested through the conjugate isomorphism map

(30) $$b \mapsto \gamma_B(\cdot).$$

For the third class of (M,R)-systems, the functional entailment is $a \in F_{S \to H}(g)$, with relational diagram

(31)

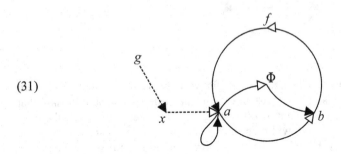

The correspondence is through the generalized inverse similarity class (*ML*: 12.12):

(32) $$A \cong H\left(A, H\left(B, H\left(A, B\right)\right)\right),$$

with

(33) $$a \mapsto \pi \circ (\cdot)^{-1}.$$

14
Therapeutics

Ne aegri quidem quia non omnes convalescunt idcirco ars nulla medicina est.

[Not all the sick recover; medicine is therefore not an art.]

      — Marcus Tullius Cicero (*c.* 45 BC)
      De Natura Deorum,
      Liber Secondus, IV(xii)

Relational Therapeutics

14.1 Three Remedies   The interaction between a host H and a symbiont S (in the most general sense when S may be external or internal to H) may be defined by the nonempty *metabolism effect* or *repair effect* of S on H : for a $g \in \kappa(S)$, either

(1) $$M_{S \to H}(g) = \upsilon(H) \cap \operatorname{ran}(g) \neq \varnothing$$

or

(2) $$F_{S \to H}(g) = \kappa(H) \cap \operatorname{ran}(g) \neq \varnothing.$$

(See Sections 10.5 and 10.6 for the definitions of the set-valued mappings $M_{S \to H} : \kappa(S) \multimap \upsilon(H)$ and $F_{S \to H} : \kappa(S) \multimap \kappa(H)$.)

This interaction may be interpreted as a perturbation of a living system H by an 'intruder' S, when g entails harmful consequences in H. I have explicated in Section 12.4 how these undesirable effect may be remedied: the treatment may be the termination, the removal, or the alteration of S. These countermeasures, which I shall explicate in this chapter, are relational models of many aspects of *therapeutics*.

A.H. Louie, *The Reflection of Life*, IFSR International Series on Systems Science and Engineering 29, DOI 10.1007/978-1-4614-6928-5_14, © Springer Science+Business Media New York 2013

14.2 Termination   The most direct termination of S is, of course, the transition

(3) $S \neq \varnothing \rightarrow S = \varnothing$.

But it does not have to be carried out to this extreme.

The Postulate of Life states that a natural system is an organism if and only if it realizes an (M,R)-system (8.30 & *ML*: 11.28). To terminate an invading organism, one may therefore simply destroy its metabolism–repair functional entailment network (as opposed to its complete material annihilation (3), from to be to not to be). Indeed, to stop a specific process $g \in \kappa(S)$ from entailing anything in H, one only needs to remove this g from $\kappa(S)$. There are, in any case, many ways to effect the transition from life to nonlife, unlike its highly nontrivial inverse process (*cf.* Section 12.13). In a host–symbiont context, the usual challenge is for the termination to be species-specific, so that in the elimination of the infective symbiont, the host does not suffer collateral damage. On the microbial level, for example, one way to terminate infective agents is by the use of a wide range of antibiotics, which serve to kill, or at least slow the growth, of the infectious microorganisms; pharmaceuticals, however, are cursed with side effects (*cf. ML*: 10.14).

Instead of reductionistic molecular therapeutics, one may think of termination treatments relationally. A material system is an organism if and only if it is closed to efficient causation (*ML*: 11.29). So to kill an organism is to open up the process closure (either by breaking up hierarchical cycles or by inserting extraneous processes that are not functionally entailed). For clef-opening strategies, one may invoke results such as

14.3 Theorem (*ML*: 13.8)   *Every   (M,R)-network   must   contain   a nonreestablishable component.*

14.4 Corollary (*ML*: 13.11)   *If a connected subnetwork of an (M,R)-network has exactly one nonreestablishable component, then that component is central.*

(A metabolism component in an (M,R)-network is *nonreestablishable* if the network does not have the capacity to replace it in the event of its absence; *cf. ML*: 13.5. A metabolism component in an (M,R)-network is *central* if its absence inhibits all activities of a connected subnetwork; *cf. ML*: 13.7.) One may even call on more abstract graph-theoretic results, such as

14.5 Theorem (The Erdős–Pósa Theorem)   *For any positive integer k, a graph either contains k disjoint cycles or else has a set of at most $4k \log k$ vertices, the deletion of which destroys all cycles.*

One must note, however, that these relational theorems are existence results, that is, one knows that 'kill-switches' exist, but there are no known algorithmic procedures to find them.

14.6 Removal   The nonempty metabolism effect (1) and repair effect (2) of S on H in their interaction involve the imminence mappings $\text{Imm}_N : \kappa(N) \multimap \kappa(N)$ for all three natural systems, $N = H, S, H \vee S$, and the compositions are

$$(4) \qquad\qquad \text{Imm}_{H \vee S} = M_{S \to H} \circ \text{Imm}_S$$

and

$$(5) \qquad\qquad \text{Imm}^2_{H \vee S} = F_{S \to H} \circ \text{Imm}_S .$$

So one sees that the set-valued mappings in the interaction, M, F, and Imm, are all predicate on the collections of the processes $\kappa(N)$ of the respective natural systems N. To remove the invader S from the host H then entails the modification of these collections so that the intersections in (1) and (2) become empty. Realized in molecular terms, some drugs combat infection by acting as biochemical blockers.

In both the termination and removal treatments, the goal is to sever the connection between H and S. In the third treatment, alteration, the intersections in (1) and (2) may stay nonempty; the strategy is to take advantage of the connection between H and S to implement an 'inverse operation' to undo the damage.

14.7 Alteration   The alteration treatment is the modification of the entailer $g \in \kappa(S)$ into g' sufficiently so that the harmful effects are neutralized and the host H returns to normalcy. It is helpful to fix ideas and consider

$$(6) \qquad\qquad g \vdash \Phi',$$

a perturbation of a repair (genetic) map Φ in the (M,R)-system H, but in view of the cyclic permutational symmetry among the three maps $\{f, \Phi, b\}$, $\Phi, \Phi' \in \kappa(H)$ may be realized as any cellular process, encompassing metabolism, repair, replication, and more.

So, for generality, let $\Phi, \Phi' \in H(Y, Z)$ where Y and Z are appropriate sets (which may themselves be hom-sets) for the roles of Φ and Φ'. (For example, when Φ and Φ' represent repair maps, they may entail metabolism maps $z \in Z = H(A, B)$ for some sets of metabolites A and B.)

It is without loss of generality to assume that the therapy g' and the original extraneous process g are processes in S, that is, that $g' \in \kappa(S)$ as well as $g \in \kappa(S)$, since one may expand the network S accordingly (by forming the join of entailment networks if necessary; *cf. ML*: 7.23). Likewise, one may also assume that g' and g have a common domain X and the same codomain $H(Y,Z)$; thus $g', g \in H(X, H(Y,Z))$. (See Section 7.18 on coproduct decomposition for the category-theoretic technicalities involved.) If g replaces Φ by $\Phi' = g(x)$ (where $x \in X$), then the ideal therapy $g \to g'$ would be a treatment process g' such that $\Phi = g'(x)$, that is, the functional entailment

$$(7) \qquad\qquad\qquad g' \vdash \Phi ,$$

recovering Φ .

In terms of the set-valued mapping $F_{S \to H} : \kappa(S) \multimap \kappa(H)$, the imminence of S on H (*cf.* Section 10.6), one has

$$(8) \qquad\qquad \Phi' \in F_{S \to H}(g) \quad \text{and} \quad \Phi \in F_{S \to H}(g') .$$

The recovery of the map Φ thus involves the 'solution' of the 'inverse inter-network imminence'

$$(9) \qquad\qquad\qquad g' = F_{S \to H}^{-1}(\Phi)$$

for g' .

Categorical Therapeutics

14.8 Image Factorization   The mapping $g' \in H(X, H(Y,Z))$, however, factors thus

(10)

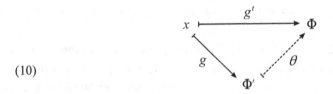

that is, into the sequential composite

(11) $$g' = \theta \circ g \,,$$

where $\theta \in H\big(H(Y,Z),H(Y,Z)\big)$ with

(12) $$\theta(\Phi') = \Phi \,.$$

The relation (12) is the conventional wisdom in therapeutics: the symptom Φ' is that which is observed, so it is treated as a material cause in search of a 'reversal therapy' θ that would materially entail the normal process Φ. The relational formulation above of the pathophysiological interactions between systems H and S has shown us the therapy that ought to be sought is the inverse functional entailment (9) instead.

14.9 Isomorphic Therapy   The pathogenetic entailment

(13) $$g \in H\big(X,H(Y,Z)\big)$$

may be curtailed in an alternate therapy if one makes use of the category-theoretic natural isomorphism

(14) $$H\big(X,H(Y,Z)\big) \cong H(X \times Y,Z)$$

(*cf.* Section 6.14). In what follows, let $x \in X$, $y,y' \in Y$, and $z,z' \in Z$.

The effect of the extraneous process g of S on the living system H is the hierarchical composite of perturbations

(15) $$\Phi \to \Phi' \text{ and } z \to z' \,,$$

that is, the mapping

(16) $$g(x) = \Phi'$$

followed by the mapping

(17) $$\Phi'(y') = z' \,.$$

The remedy

(18) $$g' \in H\big(X,H(Y,Z)\big)$$

is, therefore, (the search of) a hierarchical composite that would return H from the perturbations (15) to its normal internal processes Φ and z, that is, ideally, the mapping

(19) $g'(x) = \Phi$

followed by the mapping

(20) $\Phi(y) = z$.

Because of the isomorphism (14), g' may equivalently be considered the map

(21) $g'(x,y) = z$.

The therapy (19) followed by (20) is the hierarchical composite entailment

(22) $g' \vdash \Phi \vdash z$.

The isomorphic equivalence (21) is the entailment of the (final) final cause

(23) $g' \vdash z$.

Mathematically, the advantage of (23) over (22) is that, as a problem from which to 'solve' for the 'unknown' g', one entailment is simpler than two. (The technical procedures involve 'the inverse function theorem' from functional analysis.) Alternatively, the one equation (21) is easier to solve than the simultaneous equations (19) and (20); in particular, (21) may be solved directly, bypassing the 'intermediate' Φ. Biologically, the search for a process g' that would entail a specific gene Φ, for the purpose of replacing a mutated gene Φ' in order to restore a specific enzymatic metabolic function $z = \Phi(y)$, is *gene therapy*, the technology of which is still at the beginning of its development. The one-step inversion from the pathological $g(x,y') = z'$ back to $g'(x,y) = z$, on the other hand, offers an equivalent treatment on the physiological and biochemical level. The two treatments (22) and (23) entail the same end, differing only in their respective efficient causes of achieving it.

14.10 Reversal Therapy   The conventional 'reversal therapy' θ in relation (12) may also be helped by the natural isomorphism (14), when one considers

(24) $\theta \in H\big(H(Y,Z), H(Y,Z)\big) \cong H\big(H(Y,Z) \times Y, Z\big)$.

Thus, to solve for the 'unknown' operator θ, instead of working from

(25) $$\theta(\Phi') = \Phi,$$

one may use

(26) $$\theta(\Phi', y) = z.$$

The biological advantage of (26) over (25) is as follows. The solution of (26) for θ is the entailment

(27) $$\{\Phi', y, z\} \vdash \theta,$$

where the pathological condition Φ' is observable and the causes y and z are on lower hierarchical levels (e.g. when Φ' is a defective gene, y is a metabolite and z is an enzyme). On the other hand, the solution of (25) for θ is the entailment

(28) $$\{\Phi', \Phi\} \vdash \theta,$$

which requires the a priori knowledge of the normal process Φ in addition to the observed Φ'.

Pathogenetic Strategies and Carcinogenesis

14.11 Epigenetic Perturbation   One may speculate that the pathogenetic entailment

(29) $$g(x) = \Phi'$$

with

(30) $$g \in H\big(X, H\big(Y, H(A, B)\big)\big)$$

may also proceed through the mapping's isomorphically equivalent form

(31) $$g \in H\big(X \times Y, H(A, B)\big).$$

So the infection g, instead of entailing as the mapping (30) a direct mutation $\Phi \to \Phi'$, may equivalently entail as the mapping (31) with its primary final cause the epigenetic perturbation $f \to f'$, in the form of

(32) $g(x,y) = f'.$

(The discussion in Section 12.8 on range restriction similarly applies here: a perturbation $f \to f'$ entailed by (32) is not arbitrary, it being limited by the range of the entailer g, i.e. limited by the structure of the hom-set $H(X \times Y, H(A,B))$. So the infection strategy (32) is still on the hierarchical level of $\Phi \to \Phi'$; it is simply, through the isomorphic equivalence $H(X, H(Y, H(A,B))) \cong H(X \times Y, H(A,B))$, implementing its efficient cause on a lower level. From a set-valued mapping point of view, the entailed perturbation $f \to f'$ in (32) ultimately depends on the behaviour of $F_{S \to H} : \kappa(S) \multimap \kappa(H)$, the imminence of S on H.)

The primary epigenetic perturbation $f \to f'$ entailed by g then causes the secondary metabolic change $b \to b'$, which in turn achieves the tertiary final cause $b' \vdash \Phi'$. Stated otherwise, as the mapping (31), g institutes the causal chain

(33) $f \to f', \quad b \to b', \quad b' \vdash \Phi',$

traversing the hierarchical cycle of H to arrive at the same effect Φ'. The biological advantage (for the invader) of the multistep pathogenetic strategy (33) over the direct $g \vdash \Phi'$ as the mapping (30) is that an organism is often more vigilant in its defence on the genetic level against direct mutagenesis $\Phi \to \Phi'$. (Trivially, for the example of a eukaryotic cell, there is at least the additional physical barrier of the nuclear membrane to breach.) But the strategy (33) 'stealthily' infects on the enzymatic level and then 'hijacks' the organism's metabolism–repair–replication processes to cause the functional entailment $\vdash \Phi'$ from within.

14.12 Isomorphism Iterated   The hom-set $H(X \times Y, H(A,B))$ in (31) is such that its mappings have the hom-set $H(A,B)$ as their codomain. So the isomorphic equivalence (14) may be iterated, and one has a third description of the entailer

(34) $g \in H(X \times Y \times A, B),$

with

(35) $g(x,y,a) = b'.$

With g in this form, the primary infection strategy is the metabolic change $b \to b'$, which in turn entails the final $b' \vdash \Phi'$. (As similarly noted above, the change $b \to b'$ is not arbitrary in B : the representation (34) implicitly restricts the range of g to a proper subset of its codomain, i.e.

$$(36) \qquad g(X \times Y \times A) \subset B$$

is proper containment.) In this infection strategy, the 'stealth' is two hierarchical levels down from the final mutagenesis $\Phi \to \Phi'$.

14.13 Relational Oncology It has not escaped my notice that the category-theoretic alternate therapy I have postulated immediately suggests a possible course of treatment of the organizational disease that is cancer, for which the relational interactions that I have been exploring may be considered the invasion of a living system H by a carcinogenic S.

A scenario similar to the iterative generation of new metabolism maps in functional entailment $\vdash f$ (as I have explained in Section 12.7) can also happen here with $\vdash \Phi$. While the behaviour of the sequence

$$(37) \qquad f \to f' \to f'' \to f''' \to \cdots$$

is realized in terms of the stability of H in response to metabolic perturbations, here repeated 'exposure' to the process g of S may engender multistep genetic changes

$$(38) \qquad \Phi \to \Phi' \to \Phi'' \to \Phi''' \to \cdots,$$

each step reflecting a genetic change that cumulatively transforms a normal cell into a malignant cell. Since carcinogenesis is cumulative, the alteration treatment may target specific steps. Analogous to evolution, the reversal of the end product (e.g. the 'cure' of cancer by the retrograde and disappearance of the tumour) of the accumulation after a large number of mutational events may appear very improbable, but the reversal of each singular event is probable. In the few previous sections, we have learned ways of entailing the relational 'gene therapy' $\Phi' \to \Phi$ on various hierarchical levels.

Note that conventional cancer treatments (surgery, chemotherapy, radiation, immunotherapy, etc.) are classified in our first two remedies, termination and removal of the invader S (in this case the cancer). Relational gene therapy belongs to the third remedy, alteration of S.

14.14 Relational Antivirals   Recall (Section 13.7) that the infection by a virus S of a host cell H is a nonempty repair effect (2) of a specific form, that of $R_j \vdash M_j$ with $R_j \subset S$ and $M_j \subset H$, so that for a $g \in \kappa\left(R_j\right)$,

$$(39) \qquad F_{R_j \to M_j}\left(g\right) = \kappa\left(M_j\right) \cap \mathrm{ran}\left(g\right) \neq \varnothing .$$

Also, for the viral infection to spread, one must have the entailment closure completed by

$$(40) \qquad\qquad\qquad H \vee S \vdash g .$$

An antiviral therapy may, therefore, take advantage of these requisites. Since the repair component R_j can only entail its corresponding metabolism component M_j, instead of the complete eradication of the inter-network imminence in

$$(41) \qquad\qquad\qquad F_{S \to H}\left(g\right) = \varnothing ,$$

an antiviral process (e.g. realized molecularly as some 'antiviral drug') may simply strive for the weaker condition

$$(42) \qquad\qquad\qquad F_{R_j \to M_j}\left(g\right) = \varnothing .$$

An antiviral process may also disrupt the host+virus system $H \vee S$ sufficiently so that

$$(43) \qquad\qquad\qquad g \notin \mathrm{ran}\left(\mathrm{Imm}_{H \vee S}^{n}\right)$$

for all iterations n. Then although the original host cell H may be compromised, there will not be new copies of the virus S released to further the infection.

14.15 Disease Management   *Disease* may be defined as any condition that impairs normal function of an organism. It may be classified into two broad categories, internal dysfunction and external perturbation. An internal dysfunction may, for example, be some control error with the homeostatic mechanism or some sort of defect in the inherent organization of the organism itself. An external perturbation occurs when one organism interacts with another or simply when an organism encounters an inhospitable environment. We have seen that both classes of diseases may readily be studied relationally, through interactions between a living system H and an errant process S.

We have learned how one can study the various ways in which the metabolism–repair–replication processes in a living system can get interrupted— metabolic perturbation, genetic perturbation, epigenetic perturbation, or some combination thereof. The entailed pathophysiology is to see what the consequences would be and to see whether or not the effects are therapeutically reversible. We have discovered how relational biology may provide important insights into the kinds of causes of such physiological interruptions and the kinds of medical ways available for undoing them. Some of these treatments may lead to cures, and others only delay or slow the harm done and ameliorate the effects for as long as treatments continue. When there is cause, there is often a 'counter-cause' that can be identified, but it may be on a different hierarchical level, the remedy appearing elsewhere in the causal chain. Also, the category theory of functional entailment has shown us that both pathogenesis and therapeutics may be manifested in naturally isomorphic equivalent forms, with different degrees of subtlety in their realizations.

Parousia: A Final Reflection

14.16 Anamnesis   With this book we have taken the *second* step on our exploratory journey in relational biology.

We have begun with an exploratory introduction of the algebraic theory of set-valued mappings. Then the examination has specialized to one particular set-valued mapping, the imminence mapping

$$(44) \qquad\qquad \mathrm{Imm}_C : \mathfrak{A}C \relbar\mkern-9mu\circ \mathfrak{A}C$$

of the category **C**. The reason for this concentration is that the imminence mapping Imm_C is the very manifestation of the functional entailment structure in **C** and that a category **C** dictates the collection of (M,R)-systems (i.e. living systems) that can be formed from the **C**-objects $\mathcal{O}C$ and **C**-morphisms $\mathfrak{A}C$.

A complementary set-valued mapping, the metabolism bundle

$$(45) \qquad\qquad \mathrm{Met}_C : \mathfrak{A}C \relbar\mkern-9mu\circ \mathfrak{A}C ,$$

is the embodiment of the material entailment structure in **C**. Material entailment is generalized metabolism, and functional entailment is generalized repair. (M,R)-networks are thus generalized models of biological entities and more.

When the global set-valued mappings mapping Imm_C and Met_C are regionalized to interacting (M,R)-networks, H and S , with mutually entailing processes, they take the form of two other set-valued mappings, the repair effect $F_{S \to H}$ and metabolism effect $M_{S \to H}$. These four set-valued mappings and what they entail in biology are the *raisons d'être* of this book. Through them, and with them, and in them, one may learn (almost) everything one always wanted to know about 'What is life?' (the theme of *ML*) and how two life forms interact (the theme of this sequel). Some of this knowledge is what I have presented in the contents of this book.

14.17 Beginnings   The now-classic paper *Topology and Life* [Rashevsky 1954] is generally acknowledged as the origin of relational biology. Indeed, Nicolas Rashevsky first discussed therein the 'relational aspects' of biology. By 'relational' he meant an approach that was based on the algebraic, topological organizations of functions, as opposed to one based on the analytic, metric, mechanistic, physicochemical organizations of structures, the latter approach having heretofore dominated his subject of 'mathematical biophysics'.

Many papers from 1954 and onward, written by both Rashevsky and later Robert Rosen, mentioned the 'relational aspects', 'relational properties', 'relational approaches', etc., of biology. Rosen's first published paper [Rosen 1958] and his Rashevsky-supervised 1959 PhD thesis (in the Committee on Mathematical Biology at The University of Chicago) were both entitled *A Relational Theory of Biological Systems*. But the juxtaposition of the two words 'relational' and 'biology' (i.e. the exact phrase 'relational biology') did not appear in print until Rashevsky's 1960 paper *Contributions to Relational Biology* [Rashevsky 1960]. Rosen first used the phrase in his 1962 paper *A Note on Abstract Relational Biologies* [Rosen 1962] (the singular form appearing in the main text of the paper).

Rashevsky knew that his idea of a relational approach to biology would be a revelation in every sense of the word and, as such, something that the world might not (yet) be ready to accept (let alone to embrace). He took his leave in *Topology and Life* with a dream:

> ... the difficulties of the above approach cannot be overemphasized. This is still only a dream or vision, which may be fantastic. But if ever this dream comes true, then pure mathematics, physics, biology, and with it the social sciences will be blended into an inseparable whole.

It is now sixty years later. Where are we in this dream?

14.18 Credo   I am fortunate in having you, the reader, as a companion on our journey.

I have shown you, in our reflection of life, some of the powers of the approach that is relational biology and that many pertinent problems in biology can be better addressed this way. We have studied biogenesis, natural selection, symbiosis, pathophysiology, virology, and oncology, among others. But at the end, of course, whether my efforts have been convincing is for you to assess.

The travel from one (M,R)-system in *ML* to two interacting (M,R)-networks here in *The Reflection of Life* (henceforth denoted by the canonical symbol *RL*) has taken us to another plateau, on which the panoramic view of a new vista has opened up before us.

We will sojourn on this plateau for the present, where we may all rest, reflect on life, before pushing further. From here, we can see the next peaks to be scaled, the beyond. For now, it is enough that there is a beyond.

Visibilium omnium et invisibilium. Euouae.

Acknowledgments

I am grateful to Tim Gwinn, Roberto Poli, and Margaret Schaeken for their critical reading of the manuscript. My thanks are also due to them and to George Klir, Donald Mikulecky, and Mihai Nadin, for their continuing encouragement and motivation.

I thank George Klir for receiving this work with enthusiasm and for including it with alacrity in his *International Federation for Systems Research: International Series on Systems Science and Engineering*. My thanks are also due to Meredith Rich and Vaishali Damle, editors at Springer US, for their efforts in the publication process.

I would like to express my appreciation of my wife Margaret, without whose patience and understanding in difficult circumstances, this book would not have been written.

It is a pleasure for me to acknowledge the stalwart support of my friend I. W. Richardson.

I should have liked to have shown this book to Robert Rosen.

A. H. Louie
14 August, 2012

A.H. Louie, *The Reflection of Life*, IFSR International Series on Systems Science
and Engineering 29, DOI 10.1007/978-1-4614-6928-5,
© Springer Science+Business Media New York 2013

Bibliography

Aubin JP, Frankowska H [1990] *Set-Valued Analysis*. Birkhauser, Boston

Barrow JD, Tipler FJ [1986] *The Anthropic Cosmological Principle*. Oxford University Press, Oxford

Berge C [1963] *Topological Spaces*. Oliver & Boyd, Edinburgh

Bondy JA, Murty USR [2008] *Graph Theory*. Springer, New York

Burachik RS, Iusem AN [2008] *Set-Valued Mappings and Enlargements of Monotone Operators*. Springer, New York

de Bruijn NG, Erdős P [1951] A colour problem for infinite graphs and a problem in the theory of relations. *Indigationes Mathematicae* **13**(5): 371–373

Erdős P [1950] Some remarks on set theory. *Proceedings of the American Mathematical Society* **1**: 127–141

Erdős P, Rényi A [1959] On random graphs I. *Publicationes Mathematicae Debrecen* **6**: 290–297

Erdős P, Rényi A [1960] On the evolution of random graphs. *A Magyar Tudományos Akadémia Matematikai Kutató Intézetének Kőzleményei* **5**: 17–61

Halmos PR [1960] *Naive Set Theory*. Van Nostrand, Princeton NJ

Hardy GH [1952] *A Course of Pure Mathematics*, 10th edn. Cambridge University Press, Cambridge

Kercel SW (ed) [2005] Special issue: The work of Paul Bach-y-Rita. *Journal of Integrative Neuroscience* **4**(4)

Louie AH [1985] Categorical System Theory. In: Rosen R (ed) *Theoretical Biology and Complexity: Three Essays on the Natural Philosophy of Complex Systems*. Academic Press, Orlando FL, pp.69–163

A.H. Louie, *The Reflection of Life*, IFSR International Series on Systems Science and Engineering 29, DOI 10.1007/978-1-4614-6928-5,
© Springer Science+Business Media New York 2013

Louie AH   [2009, *ML*]   *More Than Life Itself: A Synthetic Continuation in Relational Biology.* ontos verlag, Frankfurt

Louie AH [2010] Relational biology of symbiosis. *Axiomathes* 20(4): 495–509

Louie AH [2011] Essays on More Than Life Itself. *Axiomathes* 21(3): 473–489

Louie AH, Poli R [2011] The spread of hierarchical cycles. *International Journal of General Systems* 40(3): 237–261

Poli R (ed) [2011] Special issue: Essays on *More Than Life Itself. Axiomathes* 21(3)

Rashevsky N [1954] Topology and life: In search of general mathematical principles in biology and sociology. *Bulletin of Mathematical Biophysics* 16: 317–348

Rashevsky N [1960] Contributions to relational biology. *Bulletin of Mathematical Biophysics* 22: 73–84

Rosen R [1958] A relational theory of biological systems. *Bulletin of Mathematical Biophysics* 20: 245–260

Rosen R [1962] A note on abstract relational biologies. *Bulletin of Mathematical Biophysics* 24: 31–38

Rosen R [1963] Some results in graph theory and their application to abstract relational biology. *Bulletin of Mathematical Biophysics* 25: 231–241

Rosen R [1972] Some relational cell models: the metabolism-repair systems. In: Rosen R (ed) *Foundations of Mathematical Biology*, Vol. 2. Academic Press, New York, pp.217–253

Rosen R [1991] *Life Itself.* Columbia University Press, New York

Rosen R [2006] Autobiographical reminiscences of Robert Rosen. *Axiomathes* 16(1–2): 1–23

Schrödinger E [1944] *What is Life?* Canto edition 1992. Cambridge University Press, Cambridge

Walach H, von Stillfried N (eds) [2011] Special issue: Generalizing quantum theory: approaches and applications. *Axiomathes* 21(2)

Wilkinson DM [2001] At cross purposes. *Nature* 412: 485

Index

A.H. Louie, *The Reflection of Life*, IFSR International Series on Systems Science
and Engineering 29, DOI 10.1007/978-1-4614-6928-5,
© Springer Science+Business Media New York 2013

Printed in the United States
By Bookmasters